Springer Finance

T0250732

Springer Finance

Springer Finance is a programme of books aimed at students, academics and practitioners working on increasingly technical approaches to the analysis of financial markets. It aims to cover a variety of topics, not only mathematical finance but foreign exchanges, term structure, risk management, portfolio theory, equity derivatives, and financial economics.

Ammann M., Credit Risk Valuation: Methods, Models, and Application (2001)
Back K., A Course in Derivative Securities: Introduction to Theory and Computation (2005)
Barucci E., Financial Markets Theory. Equilibrium, Efficiency and Information (2003)
Bielecki T.R. and Rutkowski M., Credit Risk: Modeling, Valuation and Hedging (2002)
Bingham N.H. and Kiesel R., Risk-Neutral Valuation: Pricing and Hedging of Financial Derivatives (1998, 2nd ed. 2004)
Brigo D. and Mercurio F., Interest Rate Models: Theory and Practice (2001)
Buff R., Uncertain Volatility Models-Theory and Application (2002)
Dana R.A. and Jeanblanc M., Financial Markets in Continuous Time (2002)
Deboeck G. and Kohonen T. (Editors), Visual Explorations in Finance with Self-Organizing Maps (1998)
Elliott R.J. and Kopp P.E., Mathematics of Financial Markets (1999, 2nd ed. 2005)
Fengler M., Semiparametric Modeling of Implied Volatility (2005)
Geman H., Madan D., Pliska S.R. and Vorst T. (Editors), Mathematical Finance–Bachelier Congress 2000 (2001)
Gundlach M., Lehrbass F. (Editors), CreditRisk$^+$ in the Banking Industry (2004)
Kellerhals B.P., Asset Pricing (2004)
Külpmann M., Irrational Exuberance Reconsidered (2004)
Kwok Y.-K., Mathematical Models of Financial Derivatives (1998)
Malliavin P. and Thalmaier A., Stochastic Calculus of Variations in Mathematical Finance (2005)
Meucci A., Risk and Asset Allocation (2005)
Pelsser A., Efficient Methods for Valuing Interest Rate Derivatives (2000)
Prigent J.-L., Weak Convergence of Financial Markets (2003)
Schmid B., Credit Risk Pricing Models (2004)
Shreve S.E., Stochastic Calculus for Finance I (2004)
Shreve S.E., Stochastic Calculus for Finance II (2004)
Yor, M., Exponential Functionals of Brownian Motion and Related Processes (2001)
Zagst R., Interest-Rate Management (2002)
Ziegler A., Incomplete Information and Heterogeneous Beliefs in Continuous-time Finance (2003)
Ziegler A., A Game Theory Analysis of Options (2004)
Zhu Y.-L., Wu X., Chern I.-L., Derivative Securities and Difference Methods (2004)

Paul Malliavin   Anton Thalmaier

Stochastic Calculus of Variations in Mathematical Finance

Paul Malliavin
Académie des Sciences
Institut de France
E-mail: *sli@ccr.jussieu.fr*

Anton Thalmaier
Département de Mathématiques
Université de Poitiers
E-mail: *anton.thalmaier@math.univ-poitiers.fr*

Mathematics Subject Classification (2000): 60H30, 60H07, 60G44, 62P20, 91B24

Library of Congress Control Number: 2005930379

ISBN-10 3-540-43431-3 Springer Berlin Heidelberg New York
ISBN-13 978-3-540-43431-3 Springer Berlin Heidelberg New York

Springer is a part of Springer Science+Business Media
springeronline.com
© Springer-Verlag Berlin Heidelberg 2006
Printed in The Netherlands

Typesetting: by the authors and TechBooks using a Springer LaTeX macro package

Cover design: *design & production,* Heidelberg

Printed on acid-free paper   SPIN: 10874794   41/TechBooks   5 4 3 2 1 0

Dedicated to Kiyosi Itô

Preface

Stochastic Calculus of Variations (or Malliavin Calculus) consists, in brief, in constructing and exploiting natural differentiable structures on abstract probability spaces; in other words, Stochastic Calculus of Variations proceeds from a merging of differential calculus and probability theory.

As optimization under a random environment is at the heart of mathematical finance, and as differential calculus is of paramount importance for the search of extrema, it is not surprising that Stochastic Calculus of Variations appears in mathematical finance. The computation of price sensitivities (or Greeks) obviously belongs to the realm of differential calculus.

Nevertheless, Stochastic Calculus of Variations was introduced relatively late in the mathematical finance literature: first in 1991 with the Ocone-Karatzas hedging formula, and soon after that, many other applications appeared in various other branches of mathematical finance; in 1999 a new impetus came from the works of P. L. Lions and his associates.

Our objective has been to write a book with complete mathematical proofs together with a relatively light conceptual load of abstract mathematics; this point of view has the drawback that often theorems are not stated under minimal hypotheses.

To faciliate applications, we emphasize, whenever possible, an approach through finite-dimensional approximation which is crucial for any kind of numerical analysis. More could have been done in numerical developments (calibrations, quantizations, etc.) and perhaps less on the geometrical approach to finance (local market stability, compartmentation by maturities of interest rate models); this bias reflects our personal background.

Chapter 1 and, to some extent, parts of Chap. 2, are the only prerequisites to reading this book; the remaining chapters should be readable independently of each other. Independence of the chapters was intended to facilitate the access to the book; sometimes however it results in closely related material being dispersed over different chapters. We hope that this inconvenience can be compensated by the extensive Index.

The authors wish to thank A. Sulem and the joint Mathematical Finance group of INRIA Rocquencourt, the Université de Marne la Vallée and Ecole Nationale des Ponts et Chaussées for the organization of an International

Symposium on the theme of our book in December 2001 (published in *Mathematical Finance*, January 2003). This Symposium was the starting point for our joint project.

Finally, we are greatly indepted to W. Schachermayer and J. Teichmann for reading a first draft of this book and for their far-reaching suggestions. Last not least, we implore the reader to send any comments on the content of this book, including errors, via email to `thalmaier@math.univ-poitiers.fr`, so that we may include them, with proper credit, in a Web page which will be created for this purpose.

Paris,   *Paul Malliavin*
April, 2005   *Anton Thalmaier*

Contents

1

Gaussian Stochastic Calculus of Variations

The Stochastic Calculus of Variations [141] has excellent basic reference articles or reference books, see for instance [40, 44, 96, 101, 144, 156, 159, 166, 169, 172, 190–193, 207]. The presentation given here will emphasize two aspects: firstly finite-dimensional approximations in view of the finite dimensionality of any set of financial data; secondly numerical constructiveness of divergence operators in view of the necessity to realize fast numerical Monte-Carlo simulations. The second point of view will be enforced through the use of *effective vector fields*.

1.1 Finite-Dimensional Gaussian Spaces, Hermite Expansion

The One-Dimensional Case

Consider the canonical Gaussian probability measure γ_1 on the real line $\mathbb{R}$ which associates to any Borel set A the mass

$$\gamma_1(A) = \frac{1}{\sqrt{2\pi}} \int_A \exp\left(-\frac{\xi^2}{2}\right) d\xi . \tag{1.1}$$

We denote by $L^2(\gamma_1)$ the Hilbert space of square-integrable functions on $\mathbb{R}$ with respect to γ_1. The monomials $\{\xi^s : s \in \mathbb{N}\}$ lie in $L^2(\gamma_1)$ and generate a dense subspace (see for instance [144], p. 6).

On dense subsets of $L^2(\gamma_1)$ there are two basic operators: the *derivative* (or *annihilation*) operator $\partial\varphi := \varphi'$ and the *creation* operator $\partial^*\varphi$, defined by

$$(\partial^*\varphi)(\xi) = -(\partial\varphi)(\xi) + \xi\varphi(\xi) . \tag{1.2}$$

Integration by parts gives the following duality formula:

$$(\partial\varphi|\psi)_{L^2(\gamma_1)} := \mathbb{E}[(\partial\varphi)\,\psi] = \int_{\mathbb{R}} (\partial\varphi)\,\psi\,d\gamma_1 = \int_{\mathbb{R}} \varphi\,(\partial^*\psi)\,d\gamma_1 = (\varphi|\partial^*\psi)_{L^2(\gamma_1)} .$$

Moreover we have the identity

$$\partial\partial^* - \partial^*\partial = 1$$

which is nothing other than the Heisenberg commutation relation; this fact explains the terminology creation, resp. annihilation operator, used in the mathematical physics literature. As the *number operator* is defined as

$$\mathcal{N} = \partial^*\partial \,, \tag{1.3}$$

we have

$$(\mathcal{N}\varphi)(\xi) = -\varphi''(\xi) + \xi\varphi'(\xi) \,.$$

Consider the sequence of Hermite polynomials given by

$$H_n(\xi) = (\partial^*)^n(1), \quad \text{i.e.,} \quad H_0(\xi) = 1, \; H_1(\xi) = \xi, \; H_2(\xi) = \xi^2 - 1, \text{ etc.}$$

Obviously H_n is a polynomial of degree n with leading term ξ^n. From the Heisenberg commutation relation we deduce that

$$\partial(\partial^*)^n - (\partial^*)^n\partial = n(\partial^*)^{n-1}.$$

Applying this identity to the constant function 1, we get

$$H_n' = nH_{n-1}, \quad \mathcal{N}H_n = nH_n \,;$$

moreover

$$\mathbb{E}[H_n H_p] = \big((\partial^*)^n 1 | H_p\big)_{L^2(\gamma_1)} = \big(1 | \partial^n H_p\big)_{L^2(\gamma_1)} = \mathbb{E}[\partial^n H_p] \,. \tag{1.4}$$

If $p < n$ the r.h.s. of (1.4) vanishes; for $p = n$ it equals $n!$. Therefore

$$\left\{ \frac{1}{\sqrt{n!}} H_n, \; n = 0, 1, \ldots \right\} \quad \text{constitutes an orthonormal basis of } L^2(\gamma_1).$$

Proposition 1.1. *Any C^∞-function φ with all its derivatives $\partial^n\varphi \in L^2(\gamma_1)$ can be represented as*

$$\varphi = \sum_{n=0}^{\infty} \frac{1}{n!} \mathbb{E}(\partial^n\varphi) H_n \,. \tag{1.5}$$

Proof. Using

$$\mathbb{E}[\partial^n\varphi] = (\partial^n\varphi \,|\, 1)_{L^2(\gamma_1)} = (\varphi \,|\, (\partial^*)^n 1)_{L^2(\gamma_1)} = \mathbb{E}[\varphi\, H_n],$$

the proof is completed by the fact that the $H_n/\sqrt{n!}$ provide an orthonormal basis of $L^2(\gamma_1)$. □

Corollary 1.2. *We have*

$$\exp\left(c\xi - \frac{1}{2}c^2\right) = \sum_{n=0}^{\infty} \frac{c^n}{n!} H_n(\xi), \quad c \in \mathbb{R}.$$

Proof. Apply (1.5) to $\varphi(\xi) := \exp(c\xi - c^2/2)$. □

The d-Dimensional Case

In the sequel, the space $\mathbb{R}^d$ is equipped with the Gaussian product measure $\gamma_d = (\gamma_1)^{\otimes d}$. Points $\xi \in \mathbb{R}^d$ are represented by their coordinates ξ^α in the standard base e_α, $\alpha = 1, \ldots, d$. The derivations (or annihilation operators) ∂_α are the partial derivatives in the direction e_α; they constitute a commuting family of operators. The creation operators ∂_α^* are now defined as

$$(\partial_\alpha^* \varphi)(\xi) := -(\partial_\alpha \varphi)(\xi) + \xi^\alpha \varphi(\xi);$$

they constitute a family of commuting operators indexed by α.

Let $\mathcal{E}$ be the set of mappings from $\{1, \ldots, d\}$ to the non-negative integers; to $\mathbf{q} \in \mathcal{E}$ we associate the following operators:

$$\partial_{\mathbf{q}} = \prod_{\alpha \in \{1, \ldots, d\}} (\partial_\alpha)^{\mathbf{q}(\alpha)}, \quad \partial_{\mathbf{q}}^* = \prod_{\alpha \in \{1, \ldots, d\}} (\partial_\alpha^*)^{\mathbf{q}(\alpha)}.$$

Duality is realized through the identities:

$$\mathbb{E}[(\partial_\alpha \varphi)\,\psi] = \mathbb{E}[\varphi\,(\partial_\alpha^* \psi)]; \quad \mathbb{E}[(\partial_{\mathbf{q}} \varphi)\,\psi] = \mathbb{E}[\varphi\,(\partial_{\mathbf{q}}^* \psi)],$$

and the commutation relationships between annihilation and creation operators are given by the Heisenberg rules:

$$\partial_\alpha^* \partial_\beta - \partial_\beta \partial_\alpha^* = \begin{cases} 1, & \text{if } \alpha = \beta \\ 0, & \text{if } \alpha \neq \beta. \end{cases}$$

The d-dimensional Hermite polynomials are indexed by $\mathcal{E}$, which means that to each $\mathbf{q} \in \mathcal{E}$ we associate

$$H_{\mathbf{q}}(\xi) := (\partial_{\mathbf{q}}^* 1)(\xi) = \prod_\alpha H_{\mathbf{q}(\alpha)}(\xi^\alpha).$$

Let $\mathbf{q}! = \prod_\alpha \mathbf{q}(\alpha)!$. Then

$$\left\{ H_{\mathbf{q}} / \sqrt{\mathbf{q}!} \right\}_{\mathbf{q} \in \mathcal{E}}$$

is an orthonormal basis of $L^2(\gamma_d)$. Defining operators ε_β on $\mathcal{E}$ by

$$(\varepsilon_\beta \mathbf{q})(\alpha) = \mathbf{q}(\alpha), \quad \text{if } \alpha \neq \beta;$$

$$(\varepsilon_\beta \mathbf{q})(\beta) = \begin{cases} \mathbf{q}(\beta) - 1, & \text{if } \mathbf{q}(\beta) > 0; \\ 0, & \text{otherwise,} \end{cases}$$

we get

$$\partial_\beta H_{\mathbf{q}} = \mathbf{q}(\beta)\, H_{\varepsilon_\beta \mathbf{q}}. \tag{1.6}$$

In generalization of the one-dimensional case given in Proposition 1.1 we now have the analogous d-dimensional result.

Proposition 1.3. *A function φ with all its partial derivatives in $L^2(\gamma_d)$ has the following representation by a series converging in $L^2(\gamma_d)$:*

$$\varphi = \sum_{\mathbf{q} \in \mathcal{E}} \frac{1}{\mathbf{q}!} \, \mathbb{E}[\partial_{\mathbf{q}} \varphi] \, H_{\mathbf{q}} \, . \tag{1.7}$$

Corollary 1.4. *For $c \in \mathbb{R}^d$ denote*

$$\|c\|^2 = \sum_\alpha (c^\alpha)^2, \quad (c \,|\, \xi) = \sum_\alpha c^\alpha \, \xi^\alpha, \quad c^{\mathbf{q}} = \prod_\alpha (c^\alpha)^{\mathbf{q}(\alpha)}.$$

Then we have

$$\exp\left((c \,|\, \xi) - \frac{1}{2}\|c\|^2 \right) = \sum_{\mathbf{q} \in \mathcal{E}} \frac{c^{\mathbf{q}}}{\mathbf{q}!} H_{\mathbf{q}}(\xi) \, . \tag{1.8}$$

In generalization of the one-dimensional case (1.3) the number operator is defined by

$$\mathcal{N} = \sum_{\alpha \in \{1,\dots,d\}} \partial_\alpha^* \partial_\alpha, \tag{1.9}$$

thus

$$(\mathcal{N}\varphi)(\xi) = \sum_{\alpha \in \{1,\dots,d\}} (-\partial_\alpha^2 \varphi + \xi^\alpha \partial_\alpha \varphi)(\xi), \quad \xi \in \mathbb{R}^d \, . \tag{1.10}$$

In particular, we get $\mathcal{N}(H_{\mathbf{q}}) = |\mathbf{q}| \, H_{\mathbf{q}}$ where $|\mathbf{q}| = \sum_\alpha \mathbf{q}(\alpha)$.

Denote by $C_b^k(\mathbb{R}^d)$ the space of k-times continuously differentiable functions on $\mathbb{R}^d$ which are bounded together with all their first k derivatives. Fix $p \geq 1$ and define a Banach type norm on $C_b^k(\mathbb{R}^d)$ by

$$\|f\|_{D_k^p}^p := \int_{\mathbb{R}^d} \left(|f|^p + \sum_{\alpha \in \{1,\dots,d\}} |\partial_\alpha f|^p \right. \tag{1.11}$$

$$\left. + \sum_{\alpha_1,\alpha_2 \in \{1,\dots,d\}} |\partial_{\alpha_1,\alpha_2}^2 f|^p + \dots + \sum_{\alpha_i \in \{1,\dots,d\}} |\partial_{\alpha_1,\dots,\alpha_k}^k f|^p \right) d\gamma_d \, .$$

A classical fact (see for instance [143]) is that the completion of $C_b^k(\mathbb{R}^d)$ in the norm $\|\cdot\|_{D_k^p}$ is the Banach space of functions for which all derivatives up to order k, computed in the sense of distributions, belong to $L^p(\gamma_d)$. We denote this completion by $D_k^p(\mathbb{R}^d)$.

Theorem 1.5. *For any $f \in C_b^2(\mathbb{R}^d)$ such that $\int f \, d\gamma_d = 0$ we have*

$$\|\mathcal{N}(f)\|_{L^2(\gamma_d)} \leq \|f\|_{D_2^2} \leq 2 \, \|\mathcal{N}(f)\|_{L^2(\gamma_d)} \, . \tag{1.12}$$

Proof. We use the expansion of f in Hermite polynomials:

$$\text{if } f = \sum_{\mathbf{q}} c_{\mathbf{q}} H_{\mathbf{q}} \quad \text{then} \quad \|f\|_{L^2(\gamma_d)}^2 = \sum_{\mathbf{q}} \mathbf{q}! \, |c_{\mathbf{q}}|^2 \, .$$

By means of (1.9) we have

$$\|\mathcal{N}(f)\|^2_{L^2(\gamma_d)} = \sum_{\mathbf{q}} |\mathbf{q}|^2 \, \mathbf{q}! \, |c_{\mathbf{q}}|^2 \, .$$

The first derivatives $\partial_\alpha f$ are computed by (1.6) and their $L^2(\gamma_d)$ norm is given by

$$\sum_{\alpha} \int_{\mathbb{R}^d} |\partial_\alpha f|^2 \, d\gamma_d = \sum_{\mathbf{q}} |c_{\mathbf{q}}|^2 \mathbf{q}! \sum_{\alpha} \mathbf{q}(\alpha) = \sum_{\mathbf{q}} |c_{\mathbf{q}}|^2 \mathbf{q}! \, |\mathbf{q}| \, .$$

The second derivatives $\partial^2_{\alpha_1,\alpha_2} f$ are computed by applying (1.6) twice and the $L^2(\gamma_d)$ norm of the second derivatives gives

$$\sum_{\alpha_1,\alpha_2} \int_{\mathbb{R}^d} |\partial^2_{\alpha_1,\alpha_2} f|^2 \, d\gamma_d = \sum_{\mathbf{q}} |c_{\mathbf{q}}|^2 \mathbf{q}! \sum_{\alpha_1,\alpha_2} \mathbf{q}(\alpha_1)\mathbf{q}(\alpha_2) = \sum_{\mathbf{q}} |c_{\mathbf{q}}|^2 \mathbf{q}! \, |\mathbf{q}|^2 \, .$$

Thus we get

$$\|f\|^2_{D^2_2} = \sum_{\mathbf{q}} |c_{\mathbf{q}}|^2 \mathbf{q}! \, (1 + |\mathbf{q}| + |\mathbf{q}|^2) \, .$$

As we supposed that $c_0 = 0$ we may assume that $|\mathbf{q}| \geq 1$. We conclude by using the inequality $x^2 < 1 + x + x^2 < 4x^2$ for $x \geq 1$. □

1.2 Wiener Space as Limit of its Dyadic Filtration

Our objective in this section is to approach the financial setting in continuous time. Strictly speaking, of course, this is a mathematical abstraction; the time series generated by the price of an asset cannot go beyond the finite amount of information in a sequence of discrete times. The advantage of continuous-time models however comes from two aspects: first it ensures stability of computations when time resolution increases, secondly models in continuous time lead to simpler and more conceptual computations than those in discrete time (simplification of Hermite expansion through iterated Itô integrals, Itô's formula, formulation of probabilistic problems in terms of PDEs).

In order to emphasize the fact that the financial reality stands in discrete time, we propose in this section a construction of the probability space underlying the Brownian motion (or the Wiener space) through a coherent sequence of discrete time approximations.

We denote by $\mathscr{W}$ the space of continuous functions $W \colon [0,1] \to \mathbb{R}$ vanishing at $t = 0$. Consider the following increasing sequence $(\mathscr{W}_s)_{s\in\mathbb{N}}$ of subspaces of $\mathscr{W}$ where $\mathscr{W}_s$ is constituted by the functions $W \in \mathscr{W}$ which are linear on each interval of the dyadic partition

$$[(k-1)2^{-s}, k2^{-s}], \quad k = 1, \ldots, 2^s \, .$$

The dimension of $\mathscr{W}_s$ is obviously 2^s, since functions in $\mathscr{W}_s$ are determined by their values assigned at $k2^{-s}$, $k = 1, \ldots, 2^s$. For each $s \in \mathbb{N}$, define a pseudo-Euclidean metric on $\mathscr{W}$ by means of

$$\|W\|_s^2 := 2^s \sum_{k=1}^{2^s} |\delta_k^s(W)|^2, \quad \delta_k^s(W) \equiv W(\delta_k^s) := W\left(\frac{k}{2^s}\right) - W\left(\frac{k-1}{2^s}\right). \tag{1.13}$$

For instance, for $\psi \in C^1([0,1]; \mathbb{R})$, we have

$$\lim_{s \to \infty} \|\psi\|_s^2 = \int_0^1 |\psi'(t)|^2 \, dt. \tag{1.14}$$

The identity $1 = 2[(\frac{1}{2})^2 + (\frac{1}{2})^2]$ induces the *compatibility principle*:

$$\|W\|_p = \|W\|_s, \quad W \in \mathscr{W}_s, \quad p \geq s. \tag{1.15}$$

On $\mathscr{W}_s$ we take the Euclidean metric defined by $\|\cdot\|_s$ and denote by γ_s^* the canonical Gaussian measure on the Euclidean space $\mathscr{W}_s$. The injection $j_s \colon \mathscr{W}_s \to \mathscr{W}$ sends the measure γ_s^* to a Borel probability measure γ_s carried by $\mathscr{W}$. Thus $\gamma_s(B) = \gamma_s^*(j_s^{-1}(B))$ for any Borel set of $\mathscr{W}$.

Let e_t be the *evaluation* at time t, that is the linear functional on $\mathscr{W}$ defined by

$$e_t \colon W \mapsto W(t),$$

and denote by $\mathscr{F}_s$ the σ-field on $\mathscr{W}$ generated by $e_{k2^{-s}}$, $k = 1, \ldots, 2^s$. By linear extrapolation between the points of the dyadic subdivision, the data $e_{k2^{-s}}$, $k = 1, \ldots, 2^s$, determine a unique element of $\mathscr{W}_s$. The algebra of Borel measurable functions which are in addition $\mathscr{F}_s$-measurable can be identified with the Borel measurable functions on $\mathscr{W}_s$.

Let $\mathscr{F}_\infty := \sigma(\cup_q \mathscr{F}_q)$. The compatibility principle (1.15) induces the following compatibility of conditional expectations:

$$\mathbb{E}^{\gamma_s}[\Phi | \mathscr{F}_q] = \mathbb{E}^{\gamma_q}[\Phi | \mathscr{F}_q], \quad \text{for } s \geq q \text{ and for all } \mathscr{F}_\infty\text{-measurable } \Phi. \tag{1.16}$$

For any $\mathscr{F}_q$-measurable function ψ, we deduce that

$$\lim_{s \to \infty} \mathbb{E}^{\gamma_s}[\psi] = \mathbb{E}^{\gamma_q}[\psi]. \tag{1.17}$$

Theorem 1.6 (Wiener). *The sequence γ_s of Borel measures on $\mathscr{W}$ converges weakly towards a probability measure γ, the Wiener measure, carried by the Hölder continuous functions of exponent $\eta < 1/2$.*

Proof. According to (1.17) we have convergence for functions which are measurable with respect to $\mathscr{F}_\infty$. As $\mathscr{F}_\infty$ generates the Borel σ-algebra of $\mathscr{W}$ for the topology of the uniform norm, it remains to prove tightness. For $\eta > 0$, a pseudo-Hölder norm on $\mathscr{W}_s$ is given by

$$\|W\|_\eta^s = 2^{-\eta s} \sup_{k \in \{1, \ldots, 2^s\}} |\delta_k^s(W)|, \quad W \in \mathscr{W}_s.$$

As each $\delta_k^s(W)$ is a Gaussian variable of variance 2^{-s}, we have

$$\gamma_s \left\{ \|W\|_\eta^s > 1 \right\} \le 2 \, \frac{2^s}{\sqrt{2\pi}} \int_{2^{s(1/2-\eta)}}^\infty \exp\left(-\frac{\xi^2}{2}\right) d\xi \le 2^s \exp\left(-\frac{2^{s(1-2\eta)}}{2}\right).$$

This estimate shows convergence of the series $\sum_s \gamma_s \left\{ \|W\|_\eta^s \ge 1 \right\}$ for $\eta < 1/2$, which implies uniform tightness of the family of measures γ_s, see Parthasarathy [175]. $\square$

The sequence of σ-subfields $\mathscr{F}_s$ provides a filtration on $\mathscr{W}$. Given $\Phi \in L^2(\mathscr{W}; \gamma)$ the conditional expectations (with respect to γ)

$$\Phi_s := \mathbb{E}^{\mathscr{F}_s}[\Phi] \tag{1.18}$$

define a martingale which converges in $L^2(\mathscr{W}; \gamma)$ to Φ.

1.3 Stroock–Sobolev Spaces of Functionals on Wiener Space

Differential calculus of functionals on the finite-dimensional Euclidean space $\mathscr{W}_s$ is defined in the usual elementary way. As we want to pass to the limit on this differential calculus, it is convenient to look upon the differential of $\psi \in C^1(\mathscr{W}_s)$ as a function defined on $[0,1]$ through the formula:

$$D_t \psi := \sum_{k=1}^{2^s} 1_{[(k-1)2^{-s}, k2^{-s}[}(t) \, \frac{\partial \psi}{\partial W(\delta_k^s)}, \quad \psi \in C^1(\mathscr{W}_s), \tag{1.19}$$

where $W(\delta_k^s)$ denotes the kth coordinate on $\mathscr{W}_s$ defined by (1.13). We denote $\delta_k^s := [(k-1)2^{-s}, k2^{-s}[$ and write $D_t \psi = \sum_{k=1}^{2^s} 1_{\delta_k^s}(t) \frac{\partial \psi}{\partial W(\delta_k^s)}$.

We have to show that (1.19) satisfies a compatibility property analogous to (1.15). To this end consider the filtered probability space constituted by the segment $[0,1]$ together with the Lebesgue measure λ, endowed with the filtration $\{\mathscr{A}_q\}$ where the sub-σ-field $\mathscr{A}_q$ is generated by $\{\delta_k^q : k = 1, \dots, 2^q\}$.

We consider the product space $\mathscr{G} := \mathscr{W} \times [0,1]$ endowed with the filtration $\mathscr{B}_s := \mathscr{F}_s \otimes \mathscr{A}_s$.

Lemma 1.7 (Cruzeiro's Compatibility Lemma [56]). *Let ϕ_q be a functional on $\mathscr{W}$ which is $\mathscr{F}_q$-measurable such that $\phi_q \in D_1^2(\mathscr{W}_q)$. Define a functional Φ_q on $\mathscr{G}$ by $\Phi_q(W, t) := (D_t \phi_q)(W)$. Consider the martingales having final values ϕ_q, Φ_q, respectively:*

$$\phi_s = \mathbb{E}^{\mathscr{F}_s}[\phi_q], \quad \Psi_s = \mathbb{E}^{\mathscr{B}_s}[\Phi_q], \quad s \le q.$$

Then $\phi_s \in D_1^2(\mathscr{W}_s)$, and furthermore,

$$(D_t \phi_s)(W) = \Psi_s(W, t). \tag{1.20}$$

Proof. It is sufficient to prove this property for $s = q-1$. The operation $\mathbb{E}^{\mathscr{F}_{q-1}}$ consists in

i) forgetting all subdivision points of the form $(2j-1)2^{-q}$,
ii) averaging on the random variables corresponding to the innovation σ-field
    $\mathcal{I}_q = \mathscr{F}_q \ominus \mathscr{F}_{q-1}$.

On the $1_{\delta_k^q}$ these operations are summarized by the formula

$$1_{\delta_k^{q-1}} = \mathbb{E}^{\mathscr{F}_q}[1_{\delta_{2k}^q} + 1_{\delta_{2k-1}^q}] .$$

Hence the compatibility principle is reduced to the following problem on $\mathbb{R}^2$. Let $\psi(x,y)$ be a C^1-function on $\mathbb{R}^2$ where (x,y) denote the standard coordinates of $\mathbb{R}^2$, and equip $\mathbb{R}^2$ with the Gaussian measure such that coordinate functions x, y become independent random variables of variance 2^{-q}; this measure is preserved by the change of variables

$$\xi = \frac{x+y}{\sqrt{2}}, \quad \eta = \frac{x-y}{\sqrt{2}} .$$

Defining

$$\theta(\xi, \eta) = \psi\left(\frac{\xi+\eta}{\sqrt{2}}, \frac{\xi-\eta}{\sqrt{2}}\right)$$

and denoting by γ_1 the normal Gaussian law on $\mathbb{R}$, we have

$$\mathbb{E}^{x+y=a}[\psi(x,y)] = \int_{-\infty}^{\infty} \theta(a, 2^{-q/2}\lambda)\,\gamma_1(d\lambda),$$

which implies the commutation

$$\frac{\partial}{\partial a}\mathbb{E}^{x+y=a} = \mathbb{E}^{x+y=a}\frac{\partial}{\partial \xi}. \quad \square$$

Definition 1.8. *We say that* $\phi \in D_1^2(\mathscr{W})$ *if* $\phi_s := \mathbb{E}^{\mathscr{F}_s}[\phi] \in D_1^2(\mathscr{W}_s)$ *for all* s, *together with the condition that the* $(\mathscr{B}_s)$-*martingale*

$$\Psi_s(W,t) := D_t(\phi_s) \tag{1.21}$$

converges in $L^2(\mathscr{G}; \gamma \otimes \lambda)$.

Remark 1.9. The fact that $\Psi_s(W,t)$ is a $\mathscr{B}_s$-martingale results from Cruzeiro's lemma.

Theorem 1.10. *There exists a natural identification between the elements of* $D_1^2(\mathscr{W})$ *and*

$$\phi \in L^2(\mathscr{W}) \quad \text{such that} \quad \sup_s \|\phi_s\|_{D_1^2(\mathscr{W}_s)} < \infty . \tag{1.22}$$

Furthermore we have:

1. *For any $\phi \in D_1^2(\mathscr{W})$ the partial derivative $D_t\phi$ is defined almost surely in (W,t).*
2. *The space $D_1^2(\mathscr{W})$ is complete with respect to the norm*

$$\|\phi\|_{D_1^2(\mathscr{W})} := \left(\mathbb{E}\left[|\phi|^2 + \int_0^1 |D_t\phi|^2 \, dt \right] \right)^{1/2}. \tag{1.23}$$

3. *Given an $F \in C^1(\mathbb{R}^n; \mathbb{R})$ with bounded first derivatives, and $\phi^1, \ldots, \phi^n$ in $D_1^2(\mathscr{W})$, then for $G(W) := F(\phi^1(W), \ldots, \phi^n(W))$ we have*

$$G \in D_1^2(\mathscr{W}), \quad D_t G = \sum_{i=1}^n \frac{\partial F}{\partial x^i} D_t\phi^i. \tag{1.24}$$

Proof. The proof proceeds in several steps:

(a) A martingale converges in L^2 if and only if its L^2 norm is bounded.
(b) An L^2 martingale converges almost surely to its L^2 limit.
(c) The space of L^2 martingales is complete.
(d) Let $\phi_s^i = \mathbb{E}^{\mathscr{F}_s}[\phi^i]$, $G_s := \mathbb{E}^{\mathscr{F}_s}[G]$. Then $G_s = F(\phi_s^1, \ldots, \phi_s^n)$, and by finite-dimensional differential calculus,

$$D_t G_s = \sum_{i=1}^n \frac{\partial F}{\partial x^i} D_t\phi_s^i,$$

which implies (1.24) by passing to the limit. □

Higher Derivatives

We consider the space $D_r^2(\mathscr{W}_s)$ of functions defined on the finite-dimensional space $\mathscr{W}_s$, for which all derivatives in the sense of distributions up to order r belong to $L^2(\gamma_s)$. The key notation is to replace integer indices of partial derivatives by continuous indices according to the following formula (written for simplicity in the case of the second derivative)

$$D_{t_1, t_2}\psi_s := \sum_{k_1, k_2 = 1}^{2^s} 1_{[(k_1-1)2^{-s}, k_1 2^{-s}[}(t_1) \, 1_{[(k_2-1)2^{-s}, k_2 2^{-s}[}(t_2) \frac{\partial\psi}{\partial\delta_{k_1}^s} \frac{\partial\psi}{\partial\delta_{k_2}^s}.$$

Cruzeiro's Compatibility Lemma 1.7 holds true also for higher derivatives which allows to extend (1.21) to (1.24) to higher derivatives. The second derivative satisfies the symmetry property $D_{t_1, t_2}\phi = D_{t_2, t_1}\phi$. More generally, derivatives of order r are symmetric functions of the indices $t_1, \ldots, t_r$.

Recall that the norm on D_1^2 is defined by (1.23):

$$\|\phi\|_{D_1^2(\mathscr{W}_s)}^2 = \mathbb{E}\left[|\phi|^2 + \int_0^1 |D_\tau\phi|^2 \, d\tau \right].$$

Definition 1.11. *The norm on* D_2^2 *is defined as*

$$\|\phi\|^2_{D_2^2(\mathscr{W}_s)} = \mathbb{E}\left[|\phi|^2 + \int_0^1 |D_\tau \phi|^2 \, d\tau + \int_0^1 \int_0^1 |D_{\tau,\lambda}\phi|^2 \, d\tau d\lambda \right]. \quad (1.25)$$

Derivatives of Cylindrical Functions

Let $t_0 \in [0,1]$ and let e_{t_0} be the corresponding evaluation map on $\mathscr{W}$ defined by $e_{t_0}(W) := W(t_0)$. If $t_0 = k_0 2^{-s_0}$ is a dyadic fraction, then for any $s \geq s_0$,

$$e_{t_0} = \sum_{k \leq k_0 2^{s-s_0}} \delta_k^s,$$

which by means of (1.19) implies that

$$D_t e_{t_0} = 1_{[0,t_0]}(t). \quad (1.26)$$

Since any $t_0 \in [0,1]$ can be approximated by dyadic fractions, the same formula is seen to hold in general. Note that, as first derivatives are constant, second order derivatives $D_{t_1,t_2} e_{t_0}$ vanish.

A *cylindrical function* Ψ is specified by points $t_1, \ldots, t_n$ in $[0,1]$ and by a differentiable function F defined on $\mathbb{R}^n$; in terms of these data the function Ψ is defined by

$$\Psi := F(e_{t_1}, \ldots, e_{t_n}).$$

From (1.24) the following formula results:

$$D_t \Psi = \sum_{i=1}^n 1_{[0,t_i[}(t) \frac{\partial F}{\partial x^i}. \quad (1.27)$$

1.4 Divergence of Vector Fields, Integration by Parts

Definition 1.12. *A* $\mathscr{B}_s$-*measurable function* Z^s *on* $\mathscr{G}$ *is called a vector field. The final value* Z^∞ *of a square-integrable* $(\mathscr{B}_s)$-*martingale* $(Z^s)_{s \geq 0}$ *on* $\mathscr{G}$ *is called an* L^2 *vector field on* $\mathscr{W}$.

For W^s fixed, the function $Z^s(W^s, \cdot)$ is defined on $[0,1]$ and constant on the intervals $](k-1)2^{-s}, k2^{-s}[$. Hence Definition 1.12 coincides with the usual definition of a vector field on $\mathbb{R}^{2^s}$; the $\{Z(W^s, k2^{-s})\}_{k=1,\ldots,2^s}$ constituting the components of the vector field.

The pairing between $\phi \in D_1^2(\mathscr{W})$ and an L^2 vector field Z^∞ is given by

$$D_{Z^\infty}\phi := \int_0^1 Z^\infty(t) \, D_t \phi \, dt = \lim_{s \to \infty} \int_0^1 Z^s(t) \, D_t \phi^s \, dt, \quad \phi^s := \mathbb{E}^{\mathscr{F}_s}(\phi); \quad (1.28)$$

the l.h.s. being an integrable random variable on $\mathscr{W}$.

Definition 1.13 (Divergence and integration by parts). *Given an L^2 vector field Z on $\mathscr{W}$, we say that Z has a divergence in L^2, denoted $\vartheta(Z)$, if $\vartheta(Z) \in L^2(\mathscr{W})$ and if*

$$\mathbb{E}[D_Z\phi] = \mathbb{E}[\phi\,\vartheta(Z)] \quad \forall \phi \in D_1^2(\mathscr{W}) . \tag{1.29}$$

Using the density of Hermite polynomials in $L^2(\mathscr{W}; \gamma_s)$, it is easy to see that if the divergence exists, it is unique.

On a finite-dimensional space the notion of divergence can be approached by an integration by parts argument within the context of classical differential calculus. For instance on $\mathbb{R}$, we may use the identity

$$\int_0^\infty Z(\xi)\,\phi'(\xi)\,\exp\left(-\frac{1}{2}\xi^2\right)d\xi = \int_0^\infty \phi(\xi)\,(\xi Z(\xi) - Z'(\xi))\,\exp\left(-\frac{1}{2}\xi^2\right)d\xi$$

which immediately gives

$$\vartheta(Z)(\xi) = \xi\,Z(\xi) - Z'(\xi).$$

This formula can be generalized to vector fields on $\mathbb{R}^d$, along with the canonical Gaussian measure, as follows

$$\vartheta(Z)(\xi) = \sum_{k=1}^d \left(\xi^k Z^k(\xi) - \frac{\partial Z^k}{\partial \xi^k}(\xi) \right) . \tag{1.30}$$

From (1.30) it is clear that computation of divergences on the Wiener space requires differentiability of vector fields; in order to reduce this differentiability to differentiability of functions as studied in Sect. 1.3, it is convenient to work with the Walsh orthonormal system of $L^2(\lambda)$ which is tailored to the filtration $(\mathscr{A}_s)$.

Denote by $\mathcal{R}$ the periodic function of period 1, which takes value 1 on the interval $[0, 1/2[$ and value -1 on $[1/2, 1[$. Recall that every non-negative integer j has a unique dyadic development

$$j = \sum_{r=0}^{+\infty} \eta_r(j)\,2^r$$

where the coefficients $\eta_r(j)$ take the value 0 or 1. Using these notations we define

$$w_j(\tau) := \prod_{r \geq 0} \mathcal{R}(\eta_r(j)\,2^r \tau) .$$

The family $(w_j)_{j \geq 0}$ constitutes an orthonormal base of $L^2([0,1]; \lambda)$. Developing $\theta \in L^2([0,1]; \lambda)$ as $\theta = \sum_{j \geq 0} \alpha_j w_j$ gives

$$\mathbb{E}^{\mathscr{A}_s}[\theta] = \sum_{0 \leq j < 2^s} \alpha_j w_j .$$

Now if Z is an L^2 vector field and W is fixed, we expand $Z(W, \tau)$ in the Walsh orthonormal system as $Z(W, \tau) = \sum_{j \geq 0} \alpha_j(W) \, w_j(\tau)$ to get $\|Z\|_{L^2}^2 = \sum_{j \geq 0} \mathbb{E}[|\alpha_j|^2]$. Finally, we define

$$D_t Z = \sum_{j \geq 0} (D_t \alpha_j) \, w_j,$$

where $Z(W, \tau) = \sum_{j \geq 0} \alpha_j(W) \, w_j(\tau)$ and

$$\|Z\|_{D_1^2}^2 := \sum_{j \geq 0} \|\alpha_j\|_{D_1^2}^2 \,. \tag{1.31}$$

Theorem 1.14. *The divergence $\vartheta(Z)$ of a vector field Z in D_1^2 exists and satisfies the Shigekawa–Nualart–Pardoux energy identity* [160, 187]:

$$\mathbb{E}\left[|\vartheta(Z)|^2\right] = \mathbb{E}\left[\int_0^1 |Z_\tau|^2 \, d\tau + \int_0^1 \int_0^1 (D_t Z_\tau)(D_\tau Z_t) \, dt \, d\tau\right]. \tag{1.32}$$

In particular the following estimate holds:

$$\mathbb{E}\left[|\vartheta(Z)|^2\right] \leq \|Z\|_{D_1^2}^2 \,. \tag{1.33}$$

Proof. First we show that estimate (1.33) is a consequence of (1.32):

$$
\begin{aligned}
\mathbb{E}\left[|\vartheta(Z)|^2\right] &= \mathbb{E}\left[\int_0^1 |Z_\tau|^2 \, d\tau + \int_0^1 \int_0^1 (D_t Z_\tau)(D_\tau Z_t) \, dt \, d\tau\right] \\
&\leq \mathbb{E}\left[\int_0^1 |Z_\tau|^2 \, d\tau + \int_0^1 \int_0^1 |D_t Z_\tau|^2 \, dt \, d\tau\right] \\
&= \|Z\|_{D_1^2}^2 \,.
\end{aligned}
$$

It remains to prove (1.32). We associate to Z the sequence

$$Z^s = \sum_{0 \leq j < 2^s} \mathbb{E}^{\mathscr{F}_s}[\alpha_j] \, w_j;$$

then Z^s may be considered as a vector field on the finite-dimensional space $\mathscr{W}^s$; therefore (1.29) can be applied to give

$$\vartheta(Z^s) = \int_0^1 \left(\dot{W}^s(\tau) Z^s(\tau) - D_\tau Z^s(\tau)\right) d\tau \,, \tag{1.34}$$

where $\dot{W}^s(\tau) = \sum_{k=1}^{2^s} W(\delta_k^s) \, 1_{](k-1)2^{-s}, \, k 2^{-s}[}(\tau)$. It should be remarked that the integral in (1.34) is the integral of an $\mathscr{F}_s$-measurable function which is constant on the subintervals of the dyadic partition of level s; integrating on each of these dyadic intervals of length 2^{-s}, we see that (1.34) writes as a finite sum, as it should be for the divergence of a vector field on $\mathbb{R}^{2^s}$. $\quad\square$

Lemma 1.15. *The divergence $\vartheta(Z^s)$ satisfies the identity (1.32).*

Proof. By means of formula (1.29) and formula (1.34) we have

$$\mathcal{J} := \mathbb{E}[\vartheta(Z^s)\,\vartheta(Z^s)]$$

$$= \mathbb{E}\left[\vartheta(Z^s)\int_0^1 \left(\dot{W}^s(\tau)Z^s(\tau) - D_\tau Z^s(\tau)\right)d\tau\right]$$

$$= \mathbb{E}\left[D_{Z^s}\int_0^1 \left(\dot{W}^s(\tau)Z^s(\tau) - D_\tau Z^s(\tau)\right)d\tau\right]$$

$$= \mathbb{E}\left[\int_0^1\int_0^1 Z^s(\tau')D_{\tau'}\left(\dot{W}^s(\tau)Z^s(\tau) - D_\tau Z^s(\tau)\right)d\tau\,d\tau'\right].$$

Computing the derivative of a product as usual we get

$$D_{\tau'}\left(\dot{W}^s(\tau)Z^s(\tau)\right) = Z^s(\tau)\left(D_{\tau'}\dot{W}^s(\tau)\right) + \dot{W}^s(\tau)\left(D_{\tau'}Z^s(\tau)\right).$$

We remark that if τ,τ' do not belong to the same dyadic interval then $D_{\tau'}\dot{W}^s(\tau) = 0$; if they do belong to the same dyadic interval the derivative is equal to 1. Note that this derivative replaces the double integral by a simple integral where we integrate on the diagonal $\tau = \tau'$; therefore

$$\mathcal{J} - \mathbb{E}\left[\int_0^1 |Z^s(\tau)|^2\,d\tau\right]$$

$$= \int_0^1\int_0^1 Z^s(\tau')\left(\dot{W}^s(\tau)(D_{\tau'}Z^s(\tau)) - D_\tau(D_{\tau'}Z^s(\tau))\right)d\tau\,d\tau',$$

where the last term has been obtained using commutation of the derivatives D_τ and $D_{\tau'}$. Introduce the vector field $Y_{\tau'}(\tau) := D_{\tau'}Z^s(\tau)$ which is considered as a vector field with respect to the variable τ, depending on the parameter τ'. Then, by means of (1.34), we have

$$\vartheta(Y_{\tau'}) = \int_0^1 \left(\dot{W}^s(\tau)(D_{\tau'}Z^s(\tau)) - D_\tau(D_{\tau'}Z^s(\tau))\right)d\tau.$$

Using this identity along with Fubini's theorem we get

$$\mathcal{J} - \mathbb{E}\left[\int_0^1 |Z^s(\tau)|^2\,d\tau\right] = \mathbb{E}\left[\int_0^1 \vartheta(Y_{\tau'})\,Z^s(\tau')\,d\tau'\right].$$

Finally, commuting $\mathbb{E}$ and the integration with respect to time, we get along with (1.29),

$$\mathcal{J} - \mathbb{E}\left[\int_0^1 |Z^s(\tau)|^2\,d\tau\right] = \int_0^1 \mathbb{E}\left[D_{Y_{\tau'}}(Z^s(\tau'))\right]d\tau'$$

$$= \int_0^1 d\tau'\,\mathbb{E}\left[\int_0^1 (D_\tau Z^s(\tau'))\,Y_{\tau'}(\tau)\,d\tau\right]. \quad \square$$

Proof (End of the proof of Theorem 1.14). As $\{Z^s\}$ is a martingale with final value Z, we have

$$\|Z^s\|_{D_1^2} \leq \|Z\|_{D_1^2}.$$

As (1.32) has been established for Z^s we can use its consequence (1.32) to obtain

$$\mathbb{E}\left[|\vartheta(Z^s)|^2\right] \leq \|Z\|_{D_1^2}^2 \ .$$

As the defining equation (1.30) is stable under conditional expectation we deduce that

$$\mathbb{E}^{\mathscr{F}_s}[\vartheta(Z^q)] = \vartheta(Z^s), \quad \forall q > s \ .$$

Therefore the sequence $\{\vartheta(Z^s)\}$ is a martingale of bounded L^2 norm, and hence converges in the L^2 norm towards a function u. By passing to the limit, u satisfies

$$\mathbb{E}[D_Z\phi] = \mathbb{E}[u\phi]$$

for any ϕ which is $\mathscr{F}_q$-measurable for some q. As these functions are dense, u must satisfy the defining relation (1.30); therefore Z has a divergence $\vartheta(Z) = u$; finally by passing to the limit, $\vartheta(Z)$ satisfies (1.32). $\square$

Proposition 1.16 (Functorial property of the divergence). *Let Z be a vector field and v be a smooth function on $\mathscr{W}$; then*

$$\vartheta(vZ) = v\vartheta(Z) - D_Z(v) \ . \tag{1.35}$$

Proof. Given a test function ϕ, then

$$D_{vZ}\phi = v\int_0^1 Z(t)\,D_t\phi\,dt = vD_Z\phi$$

and

$$\begin{aligned}\mathbb{E}[D_{vZ}\phi] = \mathbb{E}[vD_Z\phi] &= \mathbb{E}[D_Z(v\phi)] - \mathbb{E}[\phi D_Z(v)]\\ &= \mathbb{E}[\phi(v\vartheta(Z) - D_Z v)]\end{aligned}$$

which gives the claim. $\square$

Remark 1.17. The previous statement does not make precise the spaces to which each of the appearing ingredients belongs; for instance an L^2 assumption for Z and v implies a L^1 result for $D_Z v$ and the necessity of L^∞ assumptions on the test functions ϕ.

We shall use the following general result freely in the remaining part of this book.

Theorem 1.18. *For a vector field Z on $\mathscr{W}$ define*

$$\|Z\|_{D_1^p}^p := \mathbb{E}\left[\left(\int_0^1 |Z(\tau)|^2\,d\tau\right)^{p/2} + \left(\int_0^1\int_0^1 |D_t Z(\tau)|^2\,dt\,d\tau\right)^{p/2}\right] \ . \tag{1.36}$$

Then, for all $p > 1$, there exists a constant c_p such that

$$\mathbb{E}\left[|\vartheta(Z)|^p\right] \leq c_p\,\|Z\|_{D_1^p}^p \ , \tag{1.37}$$

the finiteness of the r.h.s. of (1.37) implying the existence of the divergence of Z in L^p.

Proof. See [150, 178, 212], as well as Malliavin [144], Chap. II, Theorems 6.2 and 6.2.2. □

1.5 Itô's Theory of Stochastic Integrals

The purpose of this section is to summarize without proofs some results of Itô's theory of stochastic integration. The reader interested in an exposition of Itô's theory with proofs oriented towards the Stochastic Calculus of Variations may consult [144]; see also the basic reference books [102, 149].

To stay within classical terminology, vector fields defined on Wiener space will also be called *stochastic processes*. Let $(\mathcal{N}_t)_{t \in]0,1]}$ be the filtration on $\mathscr{W}$ generated by the evaluations $\{e_\tau : \tau < t\}$. A vector field $Z(t)$ is then said to be *predictable* if $Z(t)$ is $\mathcal{N}_t$-measurable for any $t \in]0,1]$.

Proposition 1.19. *Let Z a predictable vector field in D_1^2 then*

$$D_t(Z(\tau)) = 0 \quad \lambda \otimes \lambda \text{ almost everywhere in the triangle } 0 < \tau < t < 1;$$

$$\mathbb{E}\left[|\vartheta(Z)|^2\right] = \mathbb{E}\left[\int_0^1 |Z(\tau)|^2 \, d\tau\right]. \tag{1.38}$$

Proof. The first statement results from the definition of predictability; the second claim is a consequence of formula (1.32), exploiting the fact that the integrand of the double integral $(D_t Z(\tau))(D_\tau Z(t))$ vanishes $\lambda \otimes \lambda$ everywhere on $[0,1]^2$. □

Remark 1.20. A smoothing procedure could be used to relax the D_1^2 hypothesis to an L^2 hypothesis. However we are not going to develop this point here; it will be better covered under Itô's constructive approach.

Theorem 1.21 (Itô integral). *To a given predictable L^2 vector field Z, we introduce the Itô sums*

$$\sigma_t^s(Z)(W) = \sum_{1 \le k \le t2^s} W(\delta_k^s) \, Z\left(\frac{k-1}{2^s}\right).$$

Then $\lim_{s \to \infty} \sigma_t^s(Z)$ exists and is denoted $\int_0^t Z(\tau) \, dW(\tau)$; moreover this Itô integral is a martingale:

$$\mathbb{E}^{\mathcal{N}_\sigma}\left[\int_0^t Z(\tau) \, dW(\tau)\right] = \int_0^{t \wedge \sigma} Z(\tau) \, dW(\tau),$$

and we have the energy identity:

$$\mathbb{E}\left[\left|\int_0^t Z(\tau) \, dW(\tau)\right|^2\right] = \mathbb{E}\left[\int_0^t |Z(\tau)|^2 \, d\tau\right]. \tag{1.39}$$

Proof. For instance [144], Chap. VII, Sect. 3. □

Theorem 1.22 (Itô calculus). *To given predictable L^2 processes $Z(\tau)$, $a(\tau)$ we associate a semimartingale S via*

$$S(\tau) = \int_0^\tau Z(\lambda)\, dW(\lambda) + \int_0^\tau a(\lambda)\, d\lambda \; ; \qquad (1.40)$$

in Itô's notation of stochastic differentials relation, (1.40) reads as

$$dS(\tau) = Z\, dW + a\, d\tau \; .$$

Given a deterministic C^2-function φ, the image process $\tilde{S} = \varphi \circ S$, defined by $\tilde{S}(\tau) = \varphi(S(\tau))$, is a semimartingale represented by the two processes

$$\tilde{Z}(\tau) = \varphi'(S(\tau))\, Z(\tau), \quad \tilde{a}(\tau) = \varphi'(S(\tau))\, a(\tau) + \frac{1}{2}\, \varphi''(S(\tau))\, |Z(\tau)|^2;$$

in Itô's differential notation this reads as

$$d(\varphi \circ S) = \varphi'\, dS + \frac{1}{2}\, \varphi''\, |Z|^2\, d\tau \; .$$

Proof. See [144], Chap. VII, Theorem 7.2.   □

Theorem 1.23 (Girsanov formula for changing probability measures). *Let $Z(\tau, W)$ be a bounded predictable process and consider the following semimartingale given by its Itô stochastic differential*

$$dS(W, \tau) = dW + Z(\tau, W)\, d\tau; \quad S(W, 0) = 0 \; .$$

Let γ be the Wiener measure and consider an L^2 functional $\phi(W)$ on $\mathcal{W}$. Then we have

$$\int_{\mathcal{W}} \phi(S(W, \cdot)) \exp\left(-\int_0^1 Z\, dW - \frac{1}{2} \int_0^1 |Z|^2\, d\tau\right) \gamma(dW) = \int_{\mathcal{W}} \phi(W)\, \gamma(dW) \; .$$

Proof. This is immediate by Itô calculus.   □

Theorem 1.24 (Chaos development by iterated Itô integrals). *We associate to an L^2 symmetric function $F_p \colon [0,1]^p \to \mathbb{R}$ its iterated Itô integral of order p defined as*

$$I_p(F_p)(W) = \int_0^1 dW(t_1) \int_0^{t_1} dW(t_2) \ldots \int_0^{t_{n-1}} F_p(t_1, t_2, \ldots, t_n)\, dW(t_n) \; .$$

Then the following orthogonality relations hold:

$$\mathbb{E}[I_p(F_p)\, I_q(F_q)] = 0, \quad p \neq q;$$

$$\mathbb{E}[|I_p(F_p)|^2] = \frac{1}{p!}\, \|F_p\|_{L^2([0,1]^p)}^2 \; .$$

The linear map

$$\bigoplus_{p>0} L^2_{\text{sym}}([0,1]^p) \longrightarrow L^2(\mathscr{W};\gamma), \quad F_p \mapsto \sqrt{p!}\, I_p(F_p) ,$$

is a surjective isometry onto the subspace of $L^2(\mathscr{W};\gamma)$ which is orthogonal to the function 1.

Proof. See [144], Chap. VII, Sect. 5, and Chap. VI, Theorem 2.5.3. □

Corollary 1.25 (Martingale representation through stochastic integrals). *To any process M on $\mathscr{W}$, which is a martingale with respect to the filtration $(\mathcal{N}_t)$, there exists a unique predictable process β such that*

$$M(t) = \mathbb{E}[M(1)] + \int_0^t \beta(\tau)\, dW(\tau) .$$

Proof. See [144], Chap. VII, Theorem 5.2. □

1.6 Differential and Integral Calculus in Chaos Expansion

Theorem 1.26. *A predictable L^2 vector field Z has an L^2 divergence which is given by its Itô integral:*

$$\vartheta(Z) = \int_0^1 Z(\tau)\, dW(\tau) . \tag{1.41}$$

Proof. Consider the semimartingale defined by its stochastic differential

$$dS_\varepsilon(\tau) = dW(\tau) + \varepsilon Z(S_\varepsilon(\tau),\tau)\, d\tau,$$

where ε is a small parameter. Given a test function $\phi \in D^2_1(\mathscr{W})$ we have by Theorem 1.23 (Girsanov formula)

$$\mathbb{E}\left[\phi(S_\varepsilon(W)) \exp\left(-\varepsilon \int_0^1 Z(W(\tau),\tau)\, dW(\tau) - \frac{\varepsilon^2}{2} \int_0^1 |Z(W(\tau),\tau)|^2\, d\tau\right)\right]$$
$$= \mathbb{E}[\phi(W)] .$$

Differentiating this expression with respect to ε and taking $\varepsilon = 0$, we get

$$\mathbb{E}[D_Z \phi] - \mathbb{E}\left[\phi \int_0^1 Z(\tau)\, dW(\tau)\right] = 0. □$$

Theorem 1.27 (Clark–Ocone–Karatzas formula [51, 168, 170]). *Given $\phi \in D^2_1(\mathscr{W})$, then*

$$\phi - \mathbb{E}[\phi] = \int_0^1 \mathbb{E}^{\mathcal{N}_\tau}[D_\tau \phi]\, dW(\tau) . \tag{1.42}$$

Commentary. This formula should be seen as an analogue to a basic fact in elementary differential calculus: a function of one variable can be reconstructed from the data of its derivative by computing an indefinite Riemann integral.

Proof (of the Clark–Ocone–Karatzas formula). By Corollary 1.25 there exists an L^2 process β such that

$$\phi - \mathbb{E}[\phi] = \int \beta(\tau) \, dW(\tau) .$$

For any predictable vector field Z we have by means of (1.41)

$$\mathbb{E}\left[\int D_t(\phi - \mathbb{E}[\phi]) \, Z(t) \, dt\right] = \mathbb{E}\left[(\phi - \mathbb{E}[\phi]) \int_0^1 Z(\tau) \, dW(\tau)\right]$$

$$= \mathbb{E}\left[\int \beta(\tau) \, dW(\tau) \int_0^1 Z(\tau) \, dW(\tau)\right]$$

$$= \mathbb{E}\left[\int \beta(\tau) \, Z(\tau) \, d\tau\right] ,$$

where the last identity comes from polarization of the energy identity (1.39). Therefore, we get

$$0 = \mathbb{E}\left[\int_0^1 [D_t\phi - \beta(t)] \, Z(t) \, dt\right] ,$$

and by projecting $D_t\phi - \beta(t)$ onto $\mathcal{N}_t$ we obtain

$$\mathbb{E}\left[\int_0^1 \left(\mathbb{E}^{\mathcal{N}_t}(D_t\phi) - \beta(t)\right) Z(t) \, dt\right] = 0 . \tag{1.43}$$

Since equality (1.43) holds true for any predictable vector field Z, we conclude that $\mathbb{E}^{\mathcal{N}_t}[D_t\phi] - \beta(t) = 0$. □

Theorem 1.28 (Taylor formula with a remainder term). *Given $\phi \in D_2^2(\mathcal{W})$, then*

$$\phi - \mathbb{E}[\phi] - \int_0^1 \mathbb{E}[D_t\phi] \, dW(t)$$

$$= \int_0^1 dW(t_2) \int_0^{t_2} \mathbb{E}^{\mathcal{N}_{t_1}}[D_{t_2,t_1}^2 \phi] \, dW(t_1) . \tag{1.44}$$

Proof. Fix τ and consider the scalar-valued functional $\psi_\tau := D_\tau\phi$. First applying formula (1.42) to ψ_τ, and then using the fact that $D_t(\psi_\tau) = D_{t,\tau}^2\phi$, we get

$$\psi_\tau - \mathbb{E}[D_\tau\phi] = \int_0^1 \mathbb{E}^{\mathcal{N}_t}[D_{t,\tau}^2\phi] \, dW(t) .$$

Using

$$\mathbb{E}^{\mathcal{N}_\tau}[D_\tau \phi] = \mathbb{E}^{\mathcal{N}_\tau}[\psi_\tau] = \mathbb{E}[D_\tau \phi] - \int_0^\tau \mathbb{E}^{\mathcal{N}_t}[D_{t,\tau}^2 \phi] \, dW(t) \, ,$$

we conclude by rewriting (1.42) with $\mathbb{E}^{\mathcal{N}_\tau}(D_\tau \phi)$ as given by this identity.   □

Theorem 1.29 (Stroock's differentiation formula [191, 192]). *Given a predictable vector field Z belonging to D_1^2, then*

$$D_\tau \left(\int_0^1 Z(t) \, dW(t) \right) = Z(\tau) + \int_0^1 (D_\tau Z)(t) \, dW(t) \, . \tag{1.45}$$

Proof. We remark that $D_\tau Z(t)$ is $\mathcal{N}_t$-measurable (in fact this expression vanishes for $\tau \geq t$); we prove firstly the formula for a predictable $\mathcal{B}_s$-measurable vector field Z of the type $Z = \sum_k \alpha_k 1_{\delta_k^s}$:

$$D_\tau(\vartheta(Z)) = D_\tau \left(\sum_k \alpha_k W(\delta_k^s) \right) = \sum_k \alpha_k 1_{\delta_k^s}(\tau) + \sum_k (D_\tau \alpha_k) \, W(\delta_k^s).$$

The dyadic interval containing τ then gives rise to the term $Z(\tau)$ appearing in (1.45). The result in the general case is obtained by passage to the limit.
□

Theorem 1.30 (Differential calculus in chaos expansion). *Consider a symmetric function $F_p : [0,1]^p \longrightarrow \mathbb{R}$ and denote*

$$I_p(F_p)(W) = \int_0^1 dW(t_1) \int_0^{t_1} dW(t_2) \ldots \int_0^{t_{p-1}} F_p(t_1, t_2, \ldots, t_p) \, dW(t_p)$$

the Itô iterated integral of order p. Fixing a parameter τ, we denote

$$F_p^\tau(t_1, \ldots, t_{p-1}) := F_p(\tau, t_1, \ldots, t_{p-1});$$

then

$$I_p(F_p) \in D_1^2(\mathscr{W}), \quad D_\tau(I_p(F_p)) = I_{p-1}(F_p^\tau) \, . \tag{1.46}$$

Expanding a given $\phi \in L^2$ in terms of a normalized series of iterated integrals

$$\phi - \mathbb{E}[\phi] = \sum_{p=1}^\infty \sqrt{p!} \, I_p(F_p),$$

we have $\phi \in D_1^2$ if and only if the r.h.s. of (1.47) is finite and then

$$\|\phi\|_{D_1^2}^2 = \sum_{p \geq 0} (p+1) \, \|F_p\|_{L^2([0,1])}^2 \, . \tag{1.47}$$

Proof. We establish (1.46) by recursion on the integer p. For $p = 1$, we apply (1.45) along with the fact that $D_\tau Z = 0$. Assuming the formula for all integers $p' < p$, we denote

$$Z(\lambda) = I_{p-1}\left(F_p^\lambda \prod_{1 \le s \le p-1} 1_{[0,\lambda[}(t_s)\right);$$

then

$$I_p(F_p) = \int_0^1 Z(\tau) \, dW(\tau) \, .$$

We differentiate this expression using (1.45) and the fact that $D_\tau Z = 0$ to get

$$D_\tau(I_p(F_p)) = I_{p-1}(F_p^\tau),$$

which gives the claim. □

Theorem 1.31 (Gaveau–Trauber–Skorokhod divergence formula [82]). *Given an L^2 vector field Z, for fixed τ, the $\mathbb{R}$-valued functional $Z(\tau)$ is developable in a series of iterated integrals as follows:*

$$Z(\tau) = \mathbb{E}[Z(\tau)] + \sum_{p=1}^\infty I_p(^\tau F_p) \tag{1.48}$$

where $^\tau F_p$ is a symmetric function on $[0,1]^p$ depending on the parameter τ. Denote by $^\tau_\sigma F_p$ the symmetric function of $p + 1$ variables defined by symmetrization of $G(\tau, t_1, \ldots, t_p) := {}^\tau F(t_1, \ldots, t_p)$. Then, we have

$$\vartheta(Z) = \sum_{p \ge 0} I_{p+1}(^\tau_\sigma F_p), \quad \mathbb{E}\big[|\vartheta(Z)|^2\big] \le \sum_{p \ge 0}(p+1)\frac{\|\dot{}F_p\|^2_{L^2([0,1]^{p+1})}}{p!}, \tag{1.49}$$

under the hypothesis of finiteness of the series of L^2 norms appearing in the r.h.s. of (1.49).

Proof. Note that formula (1.49) is dual to (1.47). From (1.49) there results an alternative proof of the fact that $Z \in D_1^2$ implies existence of $\vartheta(Z)$ in L^2. □

Theorem 1.32 (Stroock–Taylor formula [194]). *Let ϕ be a function which lies in $D_q^2(\mathscr{W})$ for any integer q. Then we have*

$$\phi - \mathbb{E}[\phi] = \sum_{p=1}^\infty I_p\left(\mathbb{E}\left[D_{t_1,\ldots,t_p}(\phi)\right]\right) \, . \tag{1.50}$$

Proof. We expand ϕ in chaos:

$$\phi = \sum I_p(F_p) \, .$$

The hypothesis $\phi \in D_q^2$ implies that this series is q-times differentiable; the derivative terms of order $p < q$ vanish, whereas the derivatives of terms of order $p > q$ have a vanishing expectation. The derivative of order q equals $q!\, F_q$.

Theorem 1.33. *Define the number operator (or in another terminology the Ornstein–Uhlenbeck operator) by*

$$\mathcal{N}(\phi) = \vartheta(D\phi) \ .$$

Then we have

$$\mathcal{N}(I_p(F_p)) = p\, I_p(F_p); \quad \|\mathcal{N}(\phi)\|_{L^2} \le \|\phi\|_{D_2^2},$$
$$\text{and} \quad \|\phi\|_{D_2^2} \le \sqrt{2}\, \|\mathcal{N}(\phi)\|_{L^2}, \quad \text{if } \mathbb{E}[\phi] = 0. \tag{1.51}$$

Furthermore

$$\mathcal{N}(\phi\psi) = \psi\mathcal{N}(\phi) + \phi\mathcal{N}(\psi) + 2\int_0^1 D_\tau\phi\, D_\tau\psi\, d\tau \ . \tag{1.52}$$

Consider a finite linearly independent system $h_1, \ldots, h_q$ in $L^2([0,1])$, to which we assign the $q \times q$ covariance matrix Γ defined by $\Gamma_{ij} = (h_i|h_j)_{L^2([0,1])}$. Letting $\Phi(W) = F(W(h_1), \ldots, W(h_q))$ be a generalized cylindrical function, then

$$\mathcal{N}\Phi = \Psi, \quad \Psi(W) = G(W(h_1), \ldots, W(h_q)), \quad \text{where } G = \mathcal{N}^{\{h.\}}F,$$
$$2\mathcal{N}^{\{h.\}} = \sum_{i,j} \Gamma^{ij} \left(\frac{\partial^2}{\partial\xi^i \partial\xi^j} - \xi^i \frac{\partial}{\partial\xi^j} \right), \quad (\Gamma^{ij}) = (\Gamma_{ij})^{-1} \ . \tag{1.53}$$

Proof. Following the lines of (1.12) and replacing the Hermite polynomials by iterated stochastic integrals, we get (1.51). Next we may assume that $(h_1, \ldots, h_q)$ is already an orthonormal system. Define a map $\Phi \colon \mathcal{W} \to \mathbb{R}^q$ by $W \mapsto \{W(h_i)\}$. Then $\Phi_*\gamma$ is the canonical Gauss measure ν on $\mathbb{R}^q$. The inverse image $\Phi^* \colon f \mapsto f \circ \Phi$ maps $D_2^2(\mathbb{R}^q)$ isometrically into $D_2^2(\mathcal{W})$; furthermore we have the intertwining relation $\Phi^* \circ \mathcal{N}_{\mathbb{R}^q} = \mathcal{N}_{\mathcal{W}} \circ \Phi^*$. Claim (1.53) then results immediately from formula (1.10). To prove formula (1.52), one first considers the case of generalized cylindrical functions $\varphi(W) = F(W(h_1), \ldots, W(h_q))$ and $\psi(W) = G(W(h_1), \ldots, W(h_q))$, and then concludes by means of Cruzeiro's Compatibility Lemma 1.7. $\square$

1.7 Monte-Carlo Computation of Divergence

Theorem 1.14 (or its counterpart in the chaos expansion, Theorem 1.31) provides the existence of divergence for L^2 vector fields with derivative in L^2.

For numerical purposes a pure existence theorem is not satisfactory; the intention of this section is to furnish for a large class of vector fields, the class of *effective vector fields*, a Monte-Carlo implementable procedure to compute the divergence.

A vector field U on $\mathscr{W}$ is said to be *effective* if it can be written as a finite sum of products of a smooth function by a predictable vector field:

$$U = \sum_{q=1}^{N} u_q Z_q \tag{1.54}$$

where u_q are functions in D_1^2 and where Z_q are predictable.

Theorem 1.34. *Let U be an effective vector field as given by (1.54). Then its divergence $\vartheta(U)$ exists and can be written as*

$$\vartheta(U) = \sum_{q=1}^{N} \left[u_q \sum_{k=1}^{n} \int_0^1 Z_q^k(t)\, dW^k(t) - D_{Z_q} u_q \right] .$$

Proof. As the divergence is a linear operator it is sufficient to treat the case $N = 1$. By Proposition 1.16 we can reduce ourselves to the case $u_1 = 1$; Theorem 1.26 then gives the result.

Remark 1.35. It is possible to implement the computation of an Itô integral in a Monte-Carlo simulation by using its approximation by finite sum given in Theorem 1.27.

Computation of Derivatives of Divergences

These computations will be needed later in Chap. 4. We shall first treat the case where the vector field Z is adapted. In this case the divergence equals the Itô stochastic integral.

We have the following multi-dimensional analogue of Stroock's Differentiation Theorem 1.29.

Theorem 1.36. *Let $Z \in D_1^p(\mathscr{W}^n)$ be an adapted vector field; then the corresponding Itô stochastic integral is in $D_1^p(\mathscr{W}^n)$ and we have*

$$D_{\tau,\ell}\left(\sum_{k=1}^{n} \int_0^1 Z_{t,k}\, dW^k(t) \right) = Z_{\tau,\ell} + \sum_{k=1}^{n} \int_0^1 (D_{\tau,\ell} Z_{t,k})\, dW^k(t) . \tag{1.55}$$

Corollary 1.37. *Assume that Z is a finite linear combination of adapted vector fields with smooth functions as coefficients:*

$$Z = \sum_{j=1}^{q} a_j Z^j, \quad Z^j \text{ adapted and } a_j, Z^j \in D_1^p .$$

Then by (1.35)

$$\vartheta(Z) = \sum_{j=1}^{q}(a_j\,\vartheta(Z^j) - D_{Z^j}a_j) \qquad (1.56)$$

and therefore

$$D_Y(\vartheta(Z)) = \sum_{j=1}^{q}(D_Y a_j)\,\vartheta(Z^j) + a_j\,(D_Y\vartheta(Z^j)) - D^2_{Y,Z^j}(a_j)\,, \qquad (1.57)$$

which provides a formula in explicit terms when taking $D_Y\vartheta(Z^j)$ *from* (1.56).

There exists a beautiful general commuting formula, called the *Shigekawa formula*, giving an explicit expression for $D_Y\vartheta(Z)$ without any hypothesis of adaptedness (see Shigekawa [187] or Malliavin [144] p. 58, Theorem 6.7.6). However, as this formula involves several Skorokhod integrals it is not clear how it may be used in Monte-Carlo simulations.

2

Computation of Greeks
and Integration by Parts Formulae

Typical problems in mathematical finance can be formulated in terms of PDEs (see [12, 129, 184]). In low dimensions finite element methods then provide accurate and fast numerical resolutions. The first section of this chapter quickly reviews this PDE approach.

The stochastic process describing the market is the underlying structure of the PDE approach. Stochastic analysis concepts provide a more precise light than PDEs on the structure of the problems: for instance, the classical PDE *Greeks* become *pathwise sensitivities* in the stochastic framework.

The stochastic approach to numerical analysis relies on Monte-Carlo simulations. In this context the traditional computation of Greeks appears as derivation of an empirical function, which is well known to be numerically a quite unstable procedure. The purpose of this chapter is to present the methodology of integration by parts for Monte-Carlo computation of Greeks, which from its initiation in 1999 by P. L. Lions and his associates [79, 80, 136] has stimulated the work of many other mathematicians.

2.1 PDE Option Pricing; PDEs Governing
the Evolution of Greeks

In this first section we summarize the classical mathematical finance theory of complete markets without jumps, stating its fundamental results in the language of PDEs. In the subsequent sections we shall substitute infinite-dimensional stochastic analysis for PDE theory.

The observed prices $S^i(t)$, $i = 1, \ldots, n$, of assets are driven by a diffusion operator. The prices of options are martingales with respect to the *unique risk-free measure*, see [65]. Under the risk-free measure the infinitesimal generator of the price process takes the form

$$\mathcal{L} = \frac{1}{2} \sum_{i,j=1}^{n} \alpha^{ij}(t, x) \frac{\partial^2}{\partial x^i \partial x^j} ,$$

where $\alpha = (\alpha^{ij})$ is a symmetric, positive semi-definite matrix function defined on $\mathbb{R}_+ \times \mathbb{R}^d$. The components $\alpha^{ij}(t, x)$ are known functions of (t, x) characterizing the choice of the model. For instance, in the case of a Black–Scholes model with uncorrelated assets, we have $\alpha^{ij}(t, x) = (x^i)^2$, for $j = i$, and $= 0$ otherwise.

We shall work on more general PDEs including first order terms of the type:

$$\mathscr{L} = \frac{1}{2} \sum_{i,j=1}^{n} \alpha^{ij}(t, x) \frac{\partial^2}{\partial x^i \partial x^j} + \sum_{i=1}^{n} \beta^i(t, x) \frac{\partial}{\partial x^i} \qquad (2.1)$$

with coefficients $\alpha^{ij} \colon \mathbb{R}_+ \times \mathbb{R}^d \to \mathbb{R}$ and $\beta^i \colon \mathbb{R}_+ \times \mathbb{R}^d \to \mathbb{R}$. Note that this includes also classical Black–Scholes models where under the risk-free measure first order term appear if one deals with interest rates.

Denoting by σ the (positive semi-definite) square root of the matrix $\alpha = (\alpha^{ij})$, and fixing n independent real-valued Brownian motions $W_1, \ldots, W_n$, we consider the Itô SDE

$$dS_W^i(t) = \sum_{j=1}^{n} \sigma^{ij}(t, S_W(t)) \, dW_j(t) + \beta^i(t, S_W(t)) \, dt, \quad i = 1, \ldots, n. \qquad (2.2)$$

Given a real-valued function ϕ on $\mathbb{R}^d$, we deal with European options which give the payoff $\phi(S_W^1(T), \ldots, S_W^n(T))$ at maturity time T. Assuming that the riskless interest rate is constant and equal to r, the price of this option at a time $t < T$ is given by

$$\Phi_\phi(t, x) = e^{-r(T-t)} \, \mathbb{E}[\phi(S(T)) \mid S(t) = x], \qquad (2.3)$$

if the price of the underlying asset at time t is x, i.e., if $S(t) = x$.

Theorem 2.1. *The price function satisfies the following backward heat equation*

$$\left(\frac{\partial}{\partial t} + \mathscr{L} - r \right) \Phi_\phi = 0, \quad \Phi_\phi(T, \cdot) = \phi. \qquad (2.4)$$

Sensitivities (Greeks) are infinitesimal first or second order variations of the price functional Φ_ϕ with respect to corresponding infinitesimal variations of econometric data. Sensitivities turn out to be key data for evaluating the trader's risk. The *Delta*, denoted by Δ, is defined as the *differential form* corresponding, the time being fixed, to the differential of the option price with respect to its actual position:

$$\Delta_\phi := d\left[\Phi_\phi(t, \cdot)\right] = \sum_{i=1}^{d} \Delta_\phi^i(t, x) \, dx^i, \qquad (2.5)$$

$$\Delta_\phi^i(t, x) = \frac{\partial}{\partial x^i} \Phi_\phi(t, x), \quad i = 1, \ldots, n,$$

where the operator d associates to a function f on $\mathbb{R}^n$ its differential df.

The Delta plays a key role in the computation of other sensitivities as well.

Theorem 2.2 (Prolongation theorem). *Assume that the payoff ϕ is C^1. Then the differential form $\Delta_\phi(t,x)$ satisfies the following backward matrix heat equation:*

$$\left(\frac{\partial}{\partial t} + \mathscr{L}^1 - r\right)\Delta_\phi = 0, \quad \Delta_\phi(T,\cdot) = d\phi, \tag{2.6}$$

where

$$\mathscr{L}^1 = \mathscr{L} + \sum_{i=1}^{d} M^i \frac{\partial}{\partial x^i} + \Gamma$$

and M^i, Γ denote the following matrices:

$$(M^i)^j_q = \frac{1}{2}\frac{\partial \alpha^{ij}}{\partial x^q}, \quad \Gamma^j_q = \frac{\partial \beta^j}{\partial x^q}.$$

The matrix coefficient operator $\mathscr{L}^1$ is called the prolongation of $\mathscr{L}$ and determined by the intertwining relation

$$d\mathscr{L} = \mathscr{L}^1 d. \tag{2.7}$$

Proof. Applying the differential d to the l.h.s. of (2.4), we get

$$0 = d\left(\frac{\partial}{\partial t} + \mathscr{L} - r\right)\Phi_\phi.$$

Using the commutation $d\frac{\partial}{\partial t} = \frac{\partial}{\partial t}d$ and the intertwining relation (2.7), where $\mathscr{L}^1$ is defined through this relation, we obtain (2.6). It remains to compute $\mathscr{L}^1$ in explicit terms:

$$\frac{\partial}{\partial x^q}\mathscr{L} = \frac{1}{2}\frac{\partial}{\partial x^q}\sum_{i,j=1}^{n}\alpha^{ij}\frac{\partial^2}{\partial x^i \partial x^j} + \frac{\partial}{\partial x^q}\sum_{i=1}^{n}\beta^i\frac{\partial}{\partial x^i}$$

$$= \frac{1}{2}\sum_{i,j=1}^{n}\frac{\partial}{\partial x^q}\left(\alpha^{ij}\frac{\partial^2}{\partial x^i \partial x^j}\right) + \sum_{i=1}^{n}\frac{\partial}{\partial x^q}\left(\beta^i\frac{\partial}{\partial x^i}\right)$$

$$= \mathscr{L}\frac{\partial}{\partial x^q} + \frac{1}{2}\sum_{i,j=1}^{n}\left(\frac{\partial \alpha^{ij}}{\partial x^q}\right)\frac{\partial}{\partial x^i}\frac{\partial}{\partial x^j} + \sum_{i=1}^{n}\left(\frac{\partial \beta^i}{\partial x^q}\right)\frac{\partial}{\partial x^i}$$

which proves the claim. □

Theorem 2.3 (Hedging theorem). *Keeping the hypotheses and the notation of the previous theorem, the option $\phi(S_T)$ is replicated by the following stochastic integral*

$$e^{-rT}\phi(S_W(T)) - e^{-rT}\,\mathbb{E}[\phi(S_W(T))] = \int_0^T \sum_{j=1}^{n}\gamma_W^j(t)\,dW_j(t), \tag{2.8}$$

where

$$\gamma_W^j(t) = \langle \sigma^{\cdot j}, e^{-rt} \Delta_\phi(t, S_W(t)) \rangle = e^{-rt} \sum_{i=1}^n \sigma^{ij} \Delta_\phi^i(t, S_W(t))$$

is the function obtained by coupling the vector field $\sigma^{\cdot j}$ with the differential form Δ_ϕ, multiplied by the discount factor e^{-rt}.

Proof. Since all prices are discounted with respect to the final time T, we may confine ourselves to the case of vanishing interest rate r; formula (2.8) is then received by multiplying both sides by e^{-rT}. According to Theorem 2.1, if $r = 0$, the process $M(t) = \Phi_\phi(t, S_W(t))$ is a martingale. By Corollary 1.25, we have $M(t) = M(0) + \sum_j \int_0^t \gamma^j \, dW_j$. The coefficients γ^j are then computed using Itô calculus:

$$\gamma^j \, dt = [d\Phi_\phi(t, S_W(t))] * dW_j = \sum_i \Delta_\phi^i \, (dS^i * dW_j) = \sum_i \Delta_\phi^i \, \sigma^{ij} \, dt. \quad \square$$

Remark 2.4. The importance of the Greek $\Delta_\phi(t, S)$ comes from the fact that it appears in the replication formula; this leads to a replication strategy which allows perfect hedging.

Remark 2.5. By a change of the numeraire the question of the actualized price may be treated more systematically. This problematic is however outside the scope of this book and will not be pursued here. For simplicity, we shall mainly take $r = 0$ in the sequel.

PDE Weights

A *digital European option at maturity T* is an option for which the payoff equals $\psi(S_T)$ where ψ is an indicator function, for instance, $\psi = 1_{[K,\infty[}$ where $K > 0$ is the strike prize. As $d\psi$ does not exist in the usual sense, the backward heat equation (2.6) has no obvious meaning. Let $\pi_{T \leftarrow t_0}(x_0, dx)$, $t_0 < T$, be the *fundamental solution* to the backward heat operator $\frac{\partial}{\partial t} + \mathscr{L}$, which means that

$$\left(\frac{\partial}{\partial t_0} + \mathscr{L} \right) \pi_{T \leftarrow t_0}(\cdot, dx) = 0, \quad \lim_{t_0 \to T} \pi_{T \leftarrow t_0}(x_0, dx) = \delta_{x_0},$$

where δ_{x_0} denotes the Dirac mass at the point x_0. Then, the value Φ_ϕ of the option ϕ at position (x_0, t_0) is given by

$$\Phi_\phi(t_0, x_0) = e^{-r(T-t_0)} \int_{\mathbb{R}^d} \phi(x) \, \pi_{T \leftarrow t_0}(x_0, dx) . \tag{2.9}$$

Fix $t_0 < T$ and $x_0 \in \mathbb{R}^d$. A PDE *weight* (or *elliptic weight*) associated to the vector ζ_0 is a function w_{ζ_0}, independent of ϕ, such that for any payoff function ϕ,

$$\frac{d}{d\varepsilon}\Big|_{\varepsilon=0} \Phi_\phi(t_0, x_0 + \varepsilon\zeta_0) = e^{-r(T-t_0)} \int_{\mathbb{R}^d} \phi(x) \, w_{\zeta_0}(x) \, \pi_{T \leftarrow t_0}(x_0, dx) . \tag{2.10}$$

Theorem 2.6. *Assuming ellipticity (i.e. uniform invertibility of the matrix σ) and $\sigma \in C^2$, then for any $\zeta_0 \in \mathbb{R}^d$, there exists a unique PDE weight w_{ζ_0}. Furthermore the map $\zeta_0 \mapsto w_{\zeta_0}$ is linear.*

Proof. We shall sketch an analytic proof; a detailed proof will be provided later by using probabilistic tools.

The ellipticity hypothesis implies that $\pi_{T \leftarrow t_0}(x_0, dx)$ has a density for $t_0 < T$ with respect to the Lebesgue measure,

$$\pi_{T \leftarrow t_0}(x_0, dx) = q_{T \leftarrow t_0}(x_0, x)\, dx \ ,$$

which is strictly positive, i.e., $q_{T \leftarrow t_0}(x_0, x) > 0$, and a C^1-function in the variable x_0. Then

$$w_{\zeta_0}(x) = \frac{d}{d\varepsilon}\bigg|_{\varepsilon=0} \log q_{T \leftarrow t_0}(x_0 + \varepsilon\zeta_0, x)$$

is a PDE weight. Letting $w_{\zeta_0}, w'_{\zeta_0}$ be two PDE weights, we have for any test function ϕ,

$$\int \phi(x) \left(w_{\zeta_0}(x) - w'_{\zeta_0}(x) \right) q_{T \leftarrow t_0}(x_0, x)\, dx = 0 \ .$$

As $q_{T \leftarrow t_0}(x_0, x) > 0$ for $x \in \mathbb{R}^d$, this implies that $w_{\zeta_0}(x) - w'_{\zeta_0}(x)$ almost everywhere with respect to the Lebesgue measure. Finally, exploiting uniqueness, as $w_{\zeta_0} + w_{\zeta_1}$ is a PDE weight for the vector $\zeta_0 + \zeta_1$, we deduce that $w_{\zeta_0} + w_{\zeta_1} = w_{\zeta_0+\zeta_1}$. $\square$

Example 2.7. We consider the univariate Black–Scholes model

$$dS_W(t) = S_W(t)\, dW(t)$$

and pass to logarithmic coordinates $\xi_W(t) := \log S_W(t)$. By Itô calculus, ξ_W is a solution of the SDE $d\xi_W(t) = dW - \frac{1}{2}\, dt$, and therefore

$$W(T) = \log S_W(T) - \log S_W(0) + T/2 \ .$$

The density of $S_W(T)$ is the well-known log-normal distribution:

$$p_x(y) = \frac{1}{y\sqrt{2\pi T}} \exp\left(-\frac{1}{2T} \left(\log\left(\frac{y}{x}\right) + \frac{T}{2} \right)^2 \right)$$

where $x = S_W(0)$. We conclude that

$$\Delta_\phi(x, T) = \frac{\partial}{\partial x} \int \phi(y)\, p_x(y)\, dy = \int \phi(y) \left(\frac{\partial}{\partial x} \log p_x(y) \right) p_x(y)\, dy,$$

$$\frac{\partial}{\partial x} \log p_x(y) = \frac{1}{xT} \left(\log\left(\frac{y}{x}\right) + \frac{T}{2} \right), \quad \frac{\partial}{\partial x} \log p_x(S_W(t)) = \frac{W(T)}{xT},$$

and deduce the following expression for the PDE weight w:

$$w(x) = \frac{1}{x_0} \left(\frac{1}{T} \log \frac{x}{x_0} + \frac{1}{2} \right) \ . \tag{2.11}$$

This direct approach of computation of PDE weights by integration by parts cannot always be applied because an explicit expression of the density is often lacking. In the next section we shall substitute an infinite-dimensional integration by parts on the Wiener space for the finite-dimensional integration by parts. Using effective vector fields, these infinite-dimensional integration by parts techniques will be numerically accessible by Monte-Carlo simulations.

2.2 Stochastic Flow of Diffeomorphisms; Ocone-Karatzas Hedging

We write the SDE (2.2) in vector form by introducing on $\mathbb{R}^n$ the time-dependent vector fields $A_k = (\sigma^{ik})_{1 \le i \le n}$ and $A_0 = (\beta^i)_{1 \le i \le n}$. In vectorial notation the SDE becomes

$$dS_W(t) = \sum_k A_k(t, S_W(t)) \, dW_k + A_0(t, S_W(t)) \, dt . \qquad (2.12)$$

Flows Associated to an SDE

The *flow associated to* SDE (2.12) is the map which associates $U_{t \leftarrow t_0}^W(S_0) := S_W(t)$ to $t \ge t_0$ and $S_0 \in \mathbb{R}^n$, where $S_W(\cdot)$ is the solution of (2.12) with initial condition S_0 at $t = t_0$. Since one has existence and uniqueness of solutions to an SDE with Lipschitz coefficients and given initial value, the map $U_{t \leftarrow t_0}^W$ is well-defined.

Theorem 2.8. *Assume that the maps $x \mapsto \sigma^{ij}(t,x)$, $x \mapsto \beta^i(t,x)$ are bounded and twice differentiable with bounded derivatives with respect to x, and suppose that all derivatives are continuous as functions of (t,x). Then, for any $t \ge t_0$, almost surely with respect to W, the mapping $x \mapsto U_{t \leftarrow t_0}^W(x)$ is a C^1-diffeomorphism of $\mathbb{R}^d$.*

Proof. See Nualart [159], Kunita [116], Malliavin [144] Chap. VII. □

We associate to each vector field A_k the matrix $\mathbf{A}_k$ defined by

$$(\mathbf{A}_k)_j^i = \frac{\partial A_k^i}{\partial x^j}, \quad k = 0, 1, \dots, n .$$

Define the *first order prolongation* of SDE (2.12) as the system of two SDEs:

$$dS_W(t) = \sum_k A_k(t, S_W(t)) \, dW_k + A_0(t, S_W(t)) \, dt,$$

$$d_t J_{t \leftarrow t_0}^W = \left(\sum_{k=1}^n \mathbf{A}_k(t, S_W(t)) \, dW_k(t) + \mathbf{A}_0(t, S_W(t)) \, dt \right) J_{t \leftarrow t_0}^W \qquad (2.13)$$

where $J_{t \leftarrow t_0}^W$ is a process taking its values in the real $n \times n$ matrices, satisfying the initial condition $J_{t_0 \leftarrow t_0}^W = \text{identity}$. The first order prolongation can be considered as an SDE defined on the state space $\mathbb{R}^n \oplus (\mathbb{R}^n \otimes \mathbb{R}^n)$. The second component $J_{t \leftarrow t_0}^W$ is called *Jacobian* of the flow.

The *second order prolongation* of SDE (2.12) is defined as the first order prolongation of SDE (2.13); it appears as an SDE defined on the state space $\mathbb{R}^n \oplus (\mathbb{R}^n \otimes \mathbb{R}^n) \oplus (\mathbb{R}^n \otimes \mathbb{R}^n \otimes \mathbb{R}^n)$. The second order prolongation is obtained by adjoining to the system (2.13) the third equation

$$d_t \mathcal{J}_{t \leftarrow t_0}^W = \left(\sum_{k=1}^n \mathcal{A}_k(t, S_W(t)) \, dW_k(t) + \mathcal{A}_0(t, S_W(t)) \, dt \right) \mathcal{J}_{t \leftarrow t_0}^W, \quad t \geq t_0,$$

where $[\mathcal{A}_k]_{j,s}^i := \dfrac{\partial \mathbf{A}_k}{\partial x^s} = \dfrac{\partial^2 A_k^i}{\partial x^j \partial x^s}$.

In the same way one defines third and higher order prolongations.

Theorem 2.9 (Computation of the derivative of the flow). *Fix* $\zeta_0 \in \mathbb{R}^n$, *then*

$$\left. \frac{d}{d\varepsilon} \right|_{\varepsilon=0} U_{t \leftarrow t_0}^W(S_0 + \varepsilon \zeta_0) = J_{t \leftarrow t_0}^W(\zeta_0), \quad t \geq t_0 . \tag{2.14}$$

Proof. As the SDE driving the Jacobian is the linearized equation of the SDE driving the flow, the statement of the theorem appears quite natural. For a formal proof see [144] Chap. VIII. □

Economic Meaning of the Jacobian Flow

Consider two evolutions of the same market model,

      the "random forces acting on the market" being the same.   (2.15)

This sentence is understood in the sense that sample paths of the driving Brownian motion W are the same for the two evolutions. Therefore the two evolutions *differ only with respect to their starting point at time* t_0. From a macro-econometric point of view, it is difficult to observe the effective realization of two distinct evolutions satisfying (2.15): history never repeats again; nevertheless statement (2.15) should be considered as an intellectual experiment.

Consider now a fixed evolution of the model and assume that the state S_0 of the system at time t_0 suffers an infinitesimal perturbation $S_0 \mapsto S_0 + \varepsilon \zeta_0$.

Assuming that the perturbed system satisfies (2.14), *its state at time* t *is*

$$S_W(t) + \varepsilon \zeta_W(t) + o(\varepsilon), \quad where \ \zeta_W(t) := J_{t \leftarrow t_0}^W(\zeta_0) . \tag{2.16}$$

From an econometric point of view this propagation represents the response of the economic system to the shock ζ_0, appearing during an infinitesimal interval of time at $(t_0, S_W(t_0))$.

Theorem 2.10 (Pathwise sensitivities). *To a European option of payoff ϕ at maturity T, we associate its pathwise value $\Psi(W) := \phi(S_W(T))$. Then the Malliavin derivative can be written as*

$$D_{t,k}(\Psi) = \left\langle J^W_{T\leftarrow t}(A_k)\,,\, d\phi(S_W(T)) \right\rangle . \tag{2.17}$$

The r.h.s. may be considered as a pathwise sensitivity, i.e., a "pathwise Delta". The Greek Δ^i_ϕ is given by averaging the pathwise Delta; we have

$$\Delta^i_\phi(t,x) = \sum_{k=1}^{n} \sigma_{ik}(x)\,\mathbb{E}_{t,x}[D_{t,k}(\Psi)] \tag{2.18}$$

where (σ_{ik}) denotes the inverse of the matrix (σ^{ik}), i.e., $e_i = \sum_k \sigma_{ik} A_k$.

Proof. Equation (2.17) results from (2.14). Equality (2.18) is obtained by applying Itô's formula to the r.h.s. of (2.18); the infinitesimal generator $\mathscr{L}^1$ then appears as defined in (2.6). □

The explicit shape of the Clark–Ocone–Karatzas formula (1.42) for European options can now be determined in the case of a smooth payoff function.

Theorem 2.11 (Ocone–Karatzas formula). *Assume that the payoff function ϕ is C^1, and denote*

$$\begin{aligned} \zeta^k_W(t) &:= J^W_{T\leftarrow t}\, A_k(S_W(t)), \quad t \geq t_0; \\ \beta^k_W(t) &:= \mathbb{E}^{\mathcal{N}_t}_{t_0,x_0}\left[\langle \zeta_W(t), d\phi(S_W(T))\rangle\right] . \end{aligned} \tag{2.19}$$

Then the payoff functional is represented by the following Itô integral:

$$\phi(S_W(T)) - \mathbb{E}_{t_0,x_0}[\phi(S_W(T))] = \int_{t_0}^{T} \sum_{k=1}^{d} \beta^k_W(t)\, dW_k(t) .$$

Proof. To a European option giving the payoff ϕ at maturity T, we associate the corresponding payoff functional $\Psi(W)$ defined on $\mathcal{W}$ by the formula $\Psi(W) = \phi(S_W(T))$; then we apply formula (1.42) to the functional Ψ. □

Under a *pathwise weight* we understand an $\mathbb{R}^n$-valued function $\Omega_{T\leftarrow t_0}(W)$, defined on the path space, independent of the payoff function ϕ such that

$$\mathbb{E}_{t_0,x_0}\left[\Omega^k_{T\leftarrow t_0}(W)\,\phi(S_W(T))\right] = \mathbb{E}_{t_0,x_0}[D_{t_0,k}\phi(S_W(T))], \quad k = 1,\ldots,n . \tag{2.20}$$

We remark that the l.h.s. of (2.20) determines only the conditional expectation of $\Omega^k_{T\leftarrow t_0}(W)$ under the conditioning by $S_W(T)$; there is no uniqueness of the pathwise weight $\Omega^k_{T\leftarrow t_0}(W)$. The relation between pathwise and elliptic weight is given by

$$\sum_{k=1}^{n} \sigma_{ik}(S_W(t_0))\,\mathbb{E}_{t_0,x_0}\left[\Omega^k_{T\leftarrow t_0}(W)|S_W(T) = x\right] = w_{e_i}(x) . \tag{2.21}$$

2.3 Principle of Equivalence of Instantaneous Derivatives

Integration by parts formulae have been established for vector fields Z on $\mathscr{W}$; these are maps from $\mathscr{W}$ to $L^2([t_0, T])$ which satisfy in addition a smoothness or an adaptedness condition. The directional derivative along such a vector field Z is given by

$$D_Z \psi = \sum_{k=1}^{n} \int_{t_0}^{T} D_{t,k} \psi \, Z^k(t) \, dt \; .$$

We have to deal with an *instantaneous derivation*, that is a derivation operator at a fixed time t_0, for which no formula of integration by parts is available. The strategy will be to replace the derivation at a fixed time t_0 by *pathwise smearing in time* where we differentiate on a time interval.

For given values of t_0, S_0 and ζ_0 as above, the function $\zeta_W(t) = J_{t \leftarrow t_0}^{W}(\zeta_0)$ is called the *pathwise propagation of the instantaneous derivative* at time t_0 in the direction ζ_0. This definition may be seen as the probabilistic counterpart to formula (2.6) at the level of PDEs. The main point of this pathwise propagation is the following principle.

Remark 2.12 (Principle of equivalence under the Jacobian flow). For $t \in [t_0, T]$ and $\phi \in C^1(\mathbb{R}^n)$, we have

$$D_{t_0, \zeta_W(t_0)} \big(\phi(S_W(T)) \big) = D_{t, \zeta_W(t)} \big(\phi(S_W(T)) \big) \; . \tag{2.22}$$

The meaning of (2.22) is clear from an econometric point of view: the infinitesimal perturbation $\zeta_W(t_0)$ applied at time t_0 produces the same final effect at time T as the infinitesimal perturbation $\zeta_W(t)$ applied at time t.

2.4 Pathwise Smearing for European Options

We have defined the pathwise propagation of an instantaneous derivative. From an econometric point of view this propagation represents the response of the economic system to a shock ζ_0, appearing during an infinitesimal interval of time at $(t_0, S_W(t_0))$. Such a shock is called an instantaneous perturbation. Smearing this instantaneous perturbation means "brushing out the shock effect at maturity by a continuous saving policy".

A *smearing policy of the instantaneous derivative* ζ_0 will be the data of an $\mathbb{R}^d$-valued process $\gamma^W(t) = (\gamma_1^W(t), \ldots, \gamma_n^W(t))$ satisfying the property

$$\int_{t_0}^{T} J_{T \leftarrow t}^{W} \left(\sum_{k=1}^{n} \gamma_k^W(t) A_k(S_W(t)) \right) dt = J_{T \leftarrow t_0}^{W}(\zeta_0) \; .$$

Let $J_{t_0 \leftarrow t}^{W}$ be the inverse of the matrix $J_{t \leftarrow t_0}^{W}$; then we have the following result.

Proposition 2.13. *The process γ constitutes a smearing policy of the instantaneous derivative ζ_0 if and only if*

$$\int_{t_0}^{T} J_{t_0 \leftarrow t}^{W} \left(\sum_{k=1}^{d} \gamma_k^{W}(t) A_k(S_W(t)) \right) dt = \zeta_0 \, . \tag{2.23}$$

Proof. $J_{T \leftarrow t_0}^{W} = J_{T \leftarrow t}^{W} \circ J_{t \leftarrow t_0}^{W}$. $\square$

Theorem 2.14. *A smearing policy γ of the instantaneous derivative ζ_0 defines a vector field γ^{W} on $\mathscr{W}$; denote by $\vartheta(\gamma)$ its divergence; then*

$$\frac{d}{d\varepsilon}\Big|_{\varepsilon=0} \mathbb{E}_{t_0, S_0 + \varepsilon \zeta_0} \left[\phi(S_W(T)) \right] = \mathbb{E}_{t_0, S_0} \left[\vartheta(\gamma)\, \phi(S_W(T)) \right] \, . \tag{2.24}$$

We say that the smearing policy is *adapted* if the process γ is predictable. In this case, the divergence is expressed by an Itô integral. More precisely, we get:

Theorem 2.15. [8, 38, 57, 71, 88, 195, 204] *Let γ be an adapted smearing policy, then*

$$\frac{d}{d\varepsilon}\Big|_{\varepsilon=0} \mathbb{E}_{t_0, S_0 + \varepsilon \zeta_0} \left[\phi(S_W(T)) \right] = \mathbb{E}_{t_0 S_0} \left[\phi(S_W(T)) \Omega_{T \leftarrow t_0}^{\zeta_0}(W) \right],$$

$$\Omega_{T \leftarrow t_0}^{\zeta_0}(W) := \int_{t_0}^{T} \sum_{k=1}^{d} \gamma_k(t)\, dW_k(t) \, . \tag{2.25}$$

Canonical Adapted Smearing Policy

Define $\beta_k^{W}(t)$ by the relation

$$\sum_{k=1}^{n} \beta_k^{W}(t) A_k(S_W(t)) = J_{t \leftarrow t_0}^{W}(\zeta_0) \text{ or } \sum_{k=1}^{n} \beta_k^{W}(t) J_{t_0 \leftarrow t}^{W} A_k(S_W(t)) = \zeta_0 \, .$$

Note that this relation is solvable for $\beta_k^{W}(t)$, since as a consequence of the ellipticity hypothesis, the $A_k(S_W(t))$ constitute a basis of $\mathbb{R}^d$. Remark that the processes $\beta_k^{W}(t)$ are predictable.

We choose an arbitrary predictable scalar-valued function $g^{W}(t)$ such that

$$\int_{t_0}^{T} g^{W}(t)\, dt = 1 \, .$$

Such a function will be called a *smearing gauge*. Then $\gamma = g\,\beta$ is an adapted trading policy. For instance, one may take the constant function equal to $(T - t_0)^{-1}$ as a smearing gauge, to get the *canonical adapted smearing policy*. We shall introduce in Sect. 2.6 more general choices for smearing gauges.

2.5 Examples of Computing Pathwise Weights

After circulation of [79, 80] in preprint form, methodologies have been developed in a huge number of articles related, e.g., [27–30, 32–35, 61, 87, 88, 113, 115]. A full account of these methods would surpass the limits of this book. In addition, a final judgement on various computational tricks, such as *variance reduction*, depends heavily on numerical simulations, a topic which is not touched on here.

The following section has the limited purpose of illustrating the general methodology on simple examples.

Weight for the Delta of the Univariate Black–Scholes Model

We short-hand notation by taking for initial time $t = 0$. The 1-dimensional Black–Scholes model corresponds to the geometric Brownian motion

$$dS_W(t) = \sigma\, S_W(t)\, dW(t)\,, \qquad (2.26)$$

where σ is a constant.

For a European option of payoff $\phi(S_W(T))$ the Delta is given by

$$\Delta(x,T) = \frac{d}{dx}\mathbb{E}_x\left[\phi(S_W(T))\right]\,.$$

As (2.26) is linear, it coincides with its linearization; therefore

$$J_{t\leftarrow 0}^W = \frac{S_W(t)}{S_W(0)};\quad J_{0\leftarrow t}^W = \frac{S_W(0)}{S_W(t)};\quad \beta^W(t) = \frac{1}{\sigma x}\,. \qquad (2.27)$$

We compute the weight for the Delta by taking $g(t) = 1/T$ as smearing gauge and get the smeared vector field

$$Z = \frac{1}{\sigma T x}\int_0^T D_t\, dt\,.$$

This gives

$$\Delta(x,T) = \mathbb{E}\left[D_Z\Phi(W)\right] = \mathbb{E}\left[\vartheta(Z)\,\Phi(W)\right],\quad \Phi(W) := \phi(S_W(T))\,.$$

Since Z is predictable, its divergence can be computed by the following Itô integral:

$$\vartheta(Z) := \Omega_{T\leftarrow 0}(W) = \frac{1}{\sigma T x}\int_0^T dW = \frac{W(T)}{\sigma T x}\,.$$

Taking $\sigma = 1$ gives the formula appearing on the cover of this book

$$\Delta_\phi(x,T) = \mathbb{E}_x\left[\phi(S_W(T)) \times \frac{W(T)}{xT}\right]\,. \qquad (2.28)$$

Remark 2.16. We keep the choice $\sigma = 1$ so that $\xi_W(t) = \log S_W(t)$ is a solution of the SDE $d\xi_W(t) = dW - \frac{1}{2}\,dt$. Then

$$W(T) = \log S_W(T) - \log S_W(0) + T/2,$$

and formula (2.28) coincides with formula (2.11).

Weight for the Vega of the Univariate Black–Scholes Model

By definition, the Vega is the derivative with respect to the constant volatility σ. With the replacement $\sigma \mapsto \sigma + \varepsilon$ and denoting by S_{W^ε} the corresponding solution of (2.26), we get

$$dS_{W^\varepsilon}(t) = \sigma\, S_{W^\varepsilon}(t)\, dW(t) + \varepsilon S_{W^\varepsilon}\, dW(t) . \qquad (2.29)$$

Let $\Psi_W = \frac{d}{d\varepsilon}\big|_{\varepsilon=0} S_{W^\varepsilon}$. By differentiating (2.29), Ψ_W is seen to satisfy the linear SDE

$$d\Psi_W = \sigma \Psi_W\, dW + S_W(t)\, dW, \quad \Psi_W(0) = 0 . \qquad (2.30)$$

We solve this inhomogeneous linear equation by the Lagrange method of variation of constants. Writing

$$\Psi_W(t) = J^W_{t\leftarrow 0}(u(t)) = \frac{S_W(t)}{S_W(0)} u(t) ,$$

we get by Itô calculus,

$$\begin{aligned}
d\Psi_W(t) &= \frac{S_W(t)}{S_W(0)}\, du(t) + \sigma \frac{S_W(t)}{S_W(0)} u(t)\, dW(t) + \sigma \frac{S_W(t)}{S_W(0)}\, du(t) * dW(t) \\
&= \sigma \frac{S_W(t)}{S_W(0)} u(t)\, dW(t) + S_W(t)\, dW(t)
\end{aligned}$$

which after the Lagrange simplification gives

$$du(t) + \sigma\, du * dW = x\, dW(t), \quad x = S_W(0) .$$

A consequence of the last equation is $du * dW = x\, dt$; therefore

$$u(T) = x\,(W(T) - \sigma T), \quad \Psi_W(T) = S_W(T)\,(W(T) - \sigma T) .$$

Thus, the Vega of a European option at maturity T with a payoff $\phi(S_W(T))$, where ϕ is a C^1-function, takes the form

$$V_\phi(x,T) := \mathbb{E}_x\big[\langle d\phi_{S_W(T)}, \Psi_W(T)\rangle\big] .$$

Consider the adapted vector field $Y(t) = S_W(t)/T$; then

$$\int_0^T J^W_{T\leftarrow t}(Y(t))\, dt = S_W(T)\frac{d}{dx} .$$

In terms of the function $f(W) = W(T) - \sigma T$, we get

$$V_\phi(x,T) = \mathbb{E}_x[D_{fY}\Phi(W)], \quad \Phi(W) := \phi(S_W(T)) ,$$
$$S_W(T)(W(T) - \sigma T) = x J^W_{T\leftarrow 0}(f) .$$

Consider now the adapted vector field xZ; then

$$V_\phi(x, T) = \mathbb{E}[D_{xfZ}\Phi] = \mathbb{E}[\Phi\,\vartheta(xfZ)] = \mathbb{E}\left[(xf\vartheta(Z) - D_{xZ}f)\Phi\right] .$$

By Stroock's Theorem 1.29, we get

$$D_t W(T) = D_t \int_0^T dW = 1, \quad D_{xfZ}(W(T)) = \frac{f}{\sigma} .$$

As $\vartheta(Z)$ has been computed already in (2.28):

$$\vartheta(fxZ) = \frac{W(T)^2}{\sigma T} - \frac{1}{\sigma} , \tag{2.31}$$

we conclude that

$$V_\phi(x, T) = \frac{1}{T\sigma} \, \mathbb{E}_x\left[\phi(S_W(T))\,(W(T)^2 - T)\right] . \tag{2.32}$$

2.6 Pathwise Smearing for Barrier Option

The notion of a barrier has a strong meaning in everyday economic life: bankruptcy may be thought of as touching a barrier. PDE formulations of barriers lead to boundary conditions deteriorating both regularity of solutions and efficiency of Monte-Carlo simulations (see [13, 32, 35, 83, 87, 181]).

In the first part of this section we will get estimation of Greeks by "smearing the pending accounts before bankruptcy"; the rate of smearing accelerates when bankruptcy approaches. We recover the well-known fact that Δ has the tendency to explode near the boundary.

The second part involves a more elaborated smearing by reflection on the boundary. Its Monte-Carlo implementation, which has not yet been done, seems to be relatively expensive in computing time.

Let D be a smooth open sub-domain of $\mathbb{R}^d$. Denote by T the maturity time and by τ the first exit time of $S_W(\cdot)$ from D. We consider the contingent claim given by the following functional on Wiener space $\mathscr{W}$:

$$\Psi(W) := \varphi(S_W(T))\,1_{\{T<\tau\}}(W) , \tag{2.33}$$

where $S_W(t_0) = S_0 \in D$ for some fixed $t_0 < T$ and where φ is assumed to be bounded on D.

We shall discuss in this section two different approaches to the computation of the pathwise weight for European barrier options (2.33). Recall that

$$\Delta = \frac{d}{d\varepsilon}\Big|_{\varepsilon=0} \mathbb{E}_{t_0, S_0 + \varepsilon\zeta_0}[\Psi] .$$

We consider the pathwise propagation of ζ_0 as defined by (2.13):

$$\zeta_W(t) = J^W_{t \leftarrow t_0}(\zeta_0) = \sum_{k=1}^{d} \beta^k(t)\,A_k(S_W(t)) .$$

Definition 2.17. *Let Θ denote the class of predictable processes θ such that $\theta(s) = 0$ for $s > \tau$ and such that*

$$\int_{t_0}^{T} \theta(s)\, ds = 1 \, .$$

The fact that Θ is not empty is shown later via an explicit construction.

Theorem 2.18. *The Greek Δ can be expressed by the following Itô stochastic integral: for any $\theta \in \Theta$,*

$$\Delta = \mathbb{E}\left[\Psi \, 1_{\{T < \tau\}}\, H_\zeta^\theta\right] \quad \text{where } H_\zeta^\theta := \sum_{k=1}^{d} \int_{t_0}^{T} \beta^k(s)\, \theta(s)\, dW^k(s) \, . \qquad (2.34)$$

Proof. Consider $(P_t\varphi)(S_0) := \mathbb{E}\left[\varphi(S(t))\, 1_{\{t < \tau\}}\right]$ for $t \geq t_0$. Then

$$(P_{T+t_0-t}\varphi)(S(t)), \quad t_0 \leq t \leq \min(T, \tau) \, ,$$

is a local martingale, which implies that

$$N(t) := d(P_{T+t_0-t}\varphi)_{S(t)}\, \zeta(t), \quad t_0 \leq t \leq \min(T, \tau) \, ,$$

is a local martingale as well. (Here we used the fact that the derivative of a one-parameter family of local martingales is again a local martingale.)

Define $h(t) = 1 - \int_{t_0}^{t} \theta(r)\, dr$ for $t_0 \leq t \leq T$. Since

$$N(t)h(t) - N(t_0)h(t_0) - \int_{t_0}^{t} N(r)\, dh(r) = \int_{t_0}^{t} h(r)\, dN(r) \, ,$$

we conclude that

$$d(P_{T+t_0-t}\varphi)_{S(t)}\, \zeta(t)\, h(t) - \int_{t_0}^{t} d(P_{T+t_0-r}\varphi)_{S(r)}\, \zeta(r)\, dh(r), \quad t_0 \leq t \leq T \, ,$$

is a local martingale as well. By definition, we have $dh(r) = \dot{h}(r)dr = -\theta(r)dr$. Observe now that

$$t \mapsto \int_{t_0}^{t} d(P_{T+t_0-r}\varphi)_{S(r)}\, \zeta(r)\, \theta(r)\, dr$$

$$- (P_{T+t_0-t}\varphi)(S(t)) \sum_{k=1}^{d} \int_{t_0}^{t} \theta(r)\, \beta^k(r)\, dW^k(r)$$

is a local martingale. Using $(P_{T+t_0-t}\varphi)(S(t)) = \mathbb{E}^{\mathcal{N}_t}\left[\varphi(S(T))\, 1_{\{T < \tau\}}\right]$, we conclude that

$$M(t) := d(P_{T+t_0-t}\varphi)_{S(t)}\, \zeta(t)\, h(t)$$

$$+ \mathbb{E}^{\mathcal{N}_t}\left[\varphi(S(T))\, 1_{\{T < \tau\}}\right] \sum_{k=1}^{d} \int_{t_0}^{t} \theta(r)\, \beta^k(r)\, dW^k(r)$$

is a local martingale for $t_0 \leq t \leq T' := \min(T, \tau)$. The claim now follows by comparing the expectations $\mathbb{E}[M(t_0)] = \mathbb{E}[M(T')]$. $\square$

We propose now a construction of $\theta \in \Theta$: we denote by $d(m)$ the Euclidean distance of a point $m \in D$ to ∂D, then almost surely

$$\mathcal{J}(W) = \int_0^\tau d^{-2}(S_W(r)) \, dr = +\infty \, . \tag{2.35}$$

In fact, the Green function G of D is equivalent to d at the boundary; therefore

$$\mathbb{E}[\mathcal{J}(W)] = \int_D G(S_0, x) \, dx \simeq \int_D d^{-1}(x) \, dx = +\infty \, . \tag{2.36}$$

The step from (2.36) to (2.35) requires however a delicate argument about the triviality of the tail σ-field of the Brownian motion. A more direct proof of (2.35), based on Itô's formula, can be found in Thalmaier–Wang [204], Prop. 2.3.

Define now τ_1 such that $\int_0^{\tau_1} d^{-2}(S_W(r)) \, dr = 1$; then a.s. $\tau_1 < \tau$ and we may take

$$\theta(r) = d^{-2}(S_W(r)) 1_{[0,\tau_1]}(r) \, . \tag{2.37}$$

Remark 2.19 (Interior estimates). The fact that the stopping time T comes from the exit time of the domain D has played a minor role in the previous computations. The same methodology can be applied to options which undergo a strong change in their modalities at a stopping time.

Rolling Instantaneous Derivative Pathwise Along the Barrier for a Markovian Smearing Until Maturity

We present a second methodology for computing pathwise weights for the barrier option. Unlike the first method, this second method is very specific to barrier options. It has not yet be used effectively for Monte-Carlo computations, but it has a deep mathematical significance.

We start from the point of view of intertwining partial differential operators as in Sect. 2.1.

If f satisfies Dirichlet boundary conditions on ∂D, then the *tangential component* of the differential df of f, denoted by $(df)^\rho$, vanishes on ∂D. Given the payoff function φ on D we define φ_t as in (2.4) by the backward heat semigroup

$$\partial_t \varphi_t + \Delta_1 \varphi_t = 0, \quad \varphi_T = \varphi \text{ with boundary condition } \varphi_t | \partial D = 0 \, . \tag{2.38}$$

Then $\omega_t := d\varphi_t$ satisfies the parabolic system

$$\partial_t \omega_t + \mathcal{L} \omega_t = 0, \quad \omega_T = d\varphi, \quad (\omega_t)^\rho = 0 \, . \tag{2.39}$$

The Ocone–Karatzas formula takes the form

$$\varphi - \mathbb{E}_{t_0, S_0}[\varphi] = \int_{t_0}^{T} \omega_t^k \, dW(t) \,,$$

where $\omega_t^k = \langle \omega_t \, , A_k(S(t)) \rangle$. Equation (2.39) has been solved by H. Airault [3], adopting the following methodology. Let $y_W(t)$ be *reflected Brownian motion* inside D. We take $\zeta(t) = J_{t \leftarrow t_0}^W(\zeta_0)$ if the path does not hit the boundary. "Each time" the boundary is hit at time t we take

$$\zeta(t^+) = \big(\text{orthogonal projection on the tangent hyperplane to } \partial D\big)(\zeta(t^-)) \,.$$

This definition is carried over rigorously by passing to the limit along the local time. This transport of pathwise rolling then defines a Markov process which leads to an expression for the Greek through an Itô stochastic integral:

Theorem 2.20.

$$\Delta_{\zeta(t_0)} = \langle \omega_{t_0}, \zeta(t_0) \rangle = \frac{1}{T - t_0} \, \mathbb{E}_{t_0 S_0} \left[\Psi \sum_k \int_{t_0}^{T} \beta_s^k \, dW^k(s) \right] \qquad (2.40)$$

where $\zeta(t) = \sum_{k=1}^{d} \beta^k(t) \, A_k(S(t))$.

Proof. See H. Airault [3]. □

3

Market Equilibrium and Price-Volatility Feedback Rate

This chapter is based on the article [19]. We start by describing the basic motivation behind this approach. In view of the well-known difficulty of choosing between possible models in mathematical finance, it is natural to search for approaches which are non-parametric and model-free (see for instance [77, 189] for an approach in this direction by a quite different methodology).

For assets traded at high frequency it is possible to measure the intra-day variation of historical volatility, see for example Appendix A. For a long time traders have observed general structural facts. For instance, the volatility of an asset is generally negatively correlated with the price of the same asset; this is a first order effect. The *feedback volatility rate* is a second order effect which is supposed to describe the facility for the market to absorb small variations: it appears as a sort of *liquidity index*. The effective applicability of the mathematical theorems of this chapter has to be tested in numerical studies on real ephemerides; to be statistically significant numerical experiments must show, for a well-chosen time resolution, stability of the sign of the computed feedback volatility rate.

The multivariate feedback volatility rate can be mathematically developed in an elliptic multivariate context (see [19]); however as, even at high frequency, historical cross-volatility between different assets is empirically a dubious concept, it will not be pursued here.

3.1 Natural Metric Associated to Pathwise Smearing

In Chap. 2 we realized the smearing of the instantaneous derivative in two successive operations:

(i) Construction of the pathwise transport $\zeta_W(t)$ of the instantaneous derivative.

(ii) Decomposition of $\zeta_W(t)$ with respect to the driving vector fields; the coefficients $\beta^k(W,t)$ are defined by

$$\zeta_W(t) = \sum_{k=1}^{d} \beta^k(W, t) A_k(S_W(t)),$$

where A_k denotes the kth column of the matrix $\sigma := \sqrt{\alpha}$. Note that $\sigma(\beta) = \zeta_W(t)$. The smearing is realized by the Itô stochastic integral

$$w(W) := \int_{t_0}^{T} \theta(t) \sum_{k=1}^{d} \beta^k(W, t) \, dW_k(t), \quad \theta \in \Theta, \tag{3.1}$$

where

$$\Theta = \left\{ \theta : \theta \text{ previsible such that } \int_{t_0}^{T} \theta(s) \, ds = 1 \right\}. \tag{3.2}$$

We have

$$\mathbb{E}\left[|w(W)|^2\right] = \int_{t_0}^{T} \sum_{k=1}^{d} \theta^2(t) \, |\beta_k(t)|^2 \, dt.$$

Denoting by (σ_{ik}) the inverse of the matrix (σ^{ik}), we obtain

$$\begin{aligned}
\|\zeta_W(t)\|_{\sigma^{-1}}^2 &= \sum_{i,j,k} \sigma_{ki}(S_W(t)) \, \sigma_{kj}(S_W(t)) \, \zeta_W^i(t) \, \zeta_W^j(t) \\
&= (\sigma^{-1}(\zeta_W(t)) \, | \, \sigma^{-1}(\zeta_W(t)))_{\mathbb{R}^d} \\
&= \|\sigma^{-1}(\zeta_W(t))\|_{\mathbb{R}^d}^2 \\
&= \|\beta_W(t)\|_{\mathbb{R}^d}^2.
\end{aligned}$$

This leads to the identity

$$\mathbb{E}\left[|w(W)|^2\right] = \int_{t_0}^{T} \theta^2(t) \, \|\zeta_W(t)\|_{\sigma^{-1}}^2 \, dt. \tag{3.3}$$

The qualitative meaning of (3.3) is clear; for high volatility at time t the smearing of $\zeta_W(t)$ is less expensive than for low volatility. It is thus legitimate to say that

$$\|\zeta_W(t)\|_{\sigma^{-1}(S_W(t))}$$

defines an intrinsic norm.

3.2 Price-Volatility Feedback Rate

We consider the variation of price of a single asset during a short period of the order of a few days. We use discounted values; thus actualization by the basic interest rate is not necessary.

Our basic assumption will be that the price process with respect to the risk-free measure is given by the following SDE:

$$dS_W(t) = \sigma(S_W(t))\,dW(t) - \mu(S_W(t))\,dt\,, \qquad (3.4)$$

where W is a Brownian motion and where σ, μ are unknown smooth functions describing a feedback effect of the price. For subsequent applications the case $\mu \neq 0$ will be useful, even if for the risk-free measure we have $\mu = 0$.

The pathwise sensitivity $\zeta_W(t)$ is a solution of the linearized SDE

$$d\zeta_W(t) = [\sigma'(S_W(t))\,dW(t) - \mu'(S_W(t))\,dt]\,\zeta_W(t)\,. \qquad (3.5)$$

We associate to a pathwise sensitivity $\zeta(t)$ the *rescaled variation* defined as

$$z(t) = \frac{\zeta(t)}{\sigma(t)}, \quad \text{where } \sigma(t) := \sigma(S_W(t))\,. \qquad (3.6)$$

Theorem 3.1. *The rescaled variation $z(t)$ is a differentiable function of t; its logarithmic derivative $\lambda(t)$ is called the price-volatility feedback rate function; in explicit terms we have for every $s < t$,*

$$z(t) = \exp\left(\int_s^t \lambda(S_W(\tau))\,d\tau\right) z(s), \quad \text{and} \qquad (3.7)$$

$$\lambda = -\left(\frac{1}{2}\sigma\sigma'' + \mu' - \mu\frac{\sigma'}{\sigma}\right)\,. \qquad (3.8)$$

Proof. Using Itô calculus

$$d(\sigma) = \sigma'\,(\sigma\,dW - \mu\,dt) + \frac{1}{2}\sigma''\,\sigma^2\,dt;$$

$$d\left(\frac{1}{\sigma}\right) = -\frac{\sigma'}{\sigma}\,dW - \frac{1}{2}\sigma''dt + \frac{1}{\sigma}(\sigma')^2\,dt - \frac{\sigma'}{\sigma^2}\,\mu\,dt\,.$$

Therefore the rescaled variation has the Itô differential

$$dz = \zeta\,d\left(\frac{1}{\sigma}\right) + \frac{\zeta}{\sigma}\,(\sigma'dW + \mu'\,dt) + d\zeta * d\left(\frac{1}{\sigma}\right)\,,$$

where the $*$ denotes the Itô contraction:

$$d\zeta * d\left(\frac{1}{\sigma}\right) = -\zeta\frac{(\sigma')^2}{\sigma}\,dt = -z(\sigma')^2\,dt;$$

$$\zeta\,d\left(\frac{1}{\sigma}\right) = z\left(-\sigma'\,dW - \frac{1}{2}\sigma\sigma''\,dt + (\sigma')^2\,dt + \frac{\sigma'}{\sigma}\,\mu\,dt\right)$$

$$\frac{dz}{z} = (\sigma' - \sigma')\,dW + \left(-\frac{1}{2}\sigma\sigma'' + (\sigma')^2 - (\sigma')^2 - \mu' + \mu\frac{\sigma'}{\sigma}\right)\,dt\,.$$

It is important to note that the coefficient of dW vanishes; therefore z is a differentiable function of t and

$$\lambda := \frac{\dot{z}}{z} = -\left(\frac{1}{2}\sigma\sigma'' + \mu' - \mu\frac{\sigma'}{\sigma}\right)\,. \quad \square$$

Theorem 3.2 (Meaning of the sign of λ for hedging). *Consider the risk-free process (i.e. take $\mu = 0$). If the volatility is a convex function of the price, the pathwise smearing of instantaneous derivatives decreases exponentially in time.*

Proof. The optimal smearing is given by the following minimizing problem:

$$J(T) := \inf_{\theta \in \Theta} \int_{t_0}^{T} \exp\left(2 \int_{t_0}^{t} \lambda(S_W(s))\, ds\right) \theta^2\, dt$$

where Θ is given by (3.2). If $\lambda \leq \lambda_0 < 0$, then

$$J(T) \leq \exp(2\lambda_0(T-1)), \quad T > 1 ; \tag{3.9}$$

the hedging cost of the option goes down exponentially as the maturity date increases.   □

Remark 3.3. For the Black–Scholes model we have $\lambda = 0$.

Computation of the Price-Volatility Feedback Rate in Logarithmic Coordinates

It results from Itô calculus that the rescaled variation is independent from the choice of coordinate system; therefore λ as well, which can be seen as the "appreciation rate" of the rescaled variation, is independent of the coordinate system.

Taking as coordinate the logarithm of the price and making the change of variables $x_W(t) = \log(S_W(t))$, we denote

$$a(x) = \exp(-x)\, \sigma(\exp(x)) .$$

Then, using Itô calculus, $x_W(t)$ satisfies the following SDE:

$$dx_W(t) = a(x_W(t))\, dW(t) - \frac{1}{2}a^2(x_W(t))\, dt . \tag{3.10}$$

Theorem 3.4. *The price-volatility feedback rate λ associated to (3.10) is given by*

$$\lambda = -\frac{1}{2}(a'a + aa'') . \tag{3.11}$$

Denoting by $*$ the Itô contraction, we consider the following cross-volatilities

$$dx * dx := A\, dt, \quad dA * dx := B\, dt, \quad dB * dx := C\, dt .$$

Then λ takes the form

$$\lambda = \frac{3}{8}\frac{B^2}{A^3} - \frac{1}{4}\frac{B}{A} - \frac{1}{2}\frac{C}{A^2} . \tag{3.12}$$

Proof. We apply formula (3.8) with $\mu = \frac{1}{2}a^2$. We have the Itô differential:

$$dx = a\,dW - \frac{1}{2}a^2\,dt$$

where $A = a^2$. The cross-volatility B of A and x satisfies

$$B\,dt = 2aa'\,dx * dx = 2a^3 a'\,t\,.$$

Therefore

$$aa' = \frac{B}{2a^2} = \frac{1}{2}\frac{B}{A}\,.$$

The cross-volatility C of B and x, defined by $C\,dt = dB * dx$, is given by

$$2d(aa') * dx = 2\left(aa'' + (a')^2\right)a^2\,dt = C\,dt\,.$$

But, on the other hand, we have

$$2d(aa') * dx = \frac{1}{A^2}\left(A\,(dB * dx) - B\,(dA * dx)\right) = \frac{1}{A^2}(AC - B^2)\,dt;$$

$$2aa'' = \frac{C}{A^2} - \frac{3}{2}\frac{B^2}{A^3}\,. \quad \square$$

3.3 Measurement of the Price-Volatility Feedback Rate

The volatility of the historical price process can be measured from its quadratic variation in a pathwise procedure. On the other hand, the drift driving this historical process, which is generally called the *mean risky return*, is not directly accessible from market data; only estimations using Kalman–Bucy filtering or Zakai filtering can be obtained. As a result the SDE driving the historical price process cannot be econometrically deciphered.

Theorem 3.5. *The* SDE *driving the risk-free process can be computed econometrically using a pathwise procedure.*

Proof. Its volatility is equal to the volatility of the historical price process which can be effectively measured. Its drift vanishes in the price coordinate system. $\square$

Theorem 3.6. *The price-volatility feedback rate for the risk-free process can be pathwise econometrically measured by a three-step sequence of iterative volatility measurements.*

Proof. We may use formula (3.12). First, from the observation of the price process x, the volatility A is computed; from the processes x and A the cross-volatility B is computed, and finally the cross-volatility C is computed via $C\,dt = dB * dx$. $\square$

Remark Theorem 3.6 may be seen as a result in *non-parametric Statistics.* The unknown function λ is determined from a single observation of the market evolution.

Proposition 3.7. *Keeping notation and assumptions as above, let*

$$D = \mathrm{Vol}(\log(A)), \quad E = \mathrm{Vol}(\log(D)) .$$

The price-volatility feedback rate may be expressed as

$$\lambda = -\frac{1}{4}\eta \sqrt{D}\left(\sqrt{A} + \frac{\varepsilon}{2}\sqrt{E}\right) \tag{3.13}$$

where $\varepsilon^2 = 1$ and where $\eta = 1$ if x and A are positively correlated, and $\eta = -1$ otherwise.

Remark 3.8. As the computation of volatilities is numerically more stable than the computation of cross-volatilities, formula (3.13) may be preferable to (3.12) in certain cases.

Proof. A straightforward calculation gives

$$\log A = 2\log a, \quad D = 4\left(\frac{a'}{a}\right)^2 = 4(a')^2, \quad E = 4\left(\frac{aa''}{a'}\right)^2 . \quad \square$$

The numerical exploitation of (3.12), resp. (3.13), depends on an appropriate numerical algorithm for constructing the volatility process from an empirical process. The Fourier series method [145], see Appendix A, leads by its global character to stable results for time resolutions of A in the order of $1/3$ of the resolution δ of the price process. Iterating this resolution gap we obtain that B can be calculated up to a time resolution $\delta/9$, and C up to $\delta/27$; thus the time resolution for λ lies in the range of $\delta/27$. A highly traded asset gives a time series for the price process in the range of a new quotation every 10 or 20 seconds; for such an asset the time resolution of the price-volatility feedback rate will lie in the range of minutes. See Appendix C for some results on the numerical implementation of the price-volatility feedback rate.

3.4 Market Ergodicity
and Price-Volatility Feedback Rate

To any asset two processes may be associated: the historical price process and the process corresponding to the risk-free probability measure. Any general statement on the historical price process S_t needs the choice of a model; we assume that two functions σ, μ are given such that the price process is driven by

$$dS_W(t) = \sigma(t, S_W(t))\, dW(t) - \mu(t, S_W(t))\, dt; \tag{3.14}$$

the corresponding infinitesimal generator is then

$$\mathcal{R} = \frac{1}{2}\sigma^2\frac{d^2}{dS^2} - \mu\frac{d}{dS} .$$

Definition 3.9. *A complete elliptic market is said to be ergodic if the historical price process has an invariant probability measure ρ.*

It is clear that under a long-term horizon, in the real economic world, there are no ergodic markets. On a scale of a few days however ergodicity may be observed in a market which oscillates around an equilibrium position. The purpose of this section is to prove that a negative feedback volatility rate implies ergodicity of the market.

Proposition 3.10. *Assume that $\mu = 0$. Furthermore assume that there exists $\delta > 0$ such that the price-volatility feedback rate associated to (3.14) satisfies for all t*

$$\lambda(t) < -\delta . \tag{3.15}$$

Then the market has no remote memory (that is $z_W(t) \to 0$ as $t \to +\infty$). More precisely, we have the estimate

$$|z_W(t)| \le \exp\left(-\delta(t - t_0)\right)|z_W(t_0)|, \quad \forall t > t_0 . \tag{3.16}$$

Proof. The proof results from (3.7). □

Theorem 3.11. *We keep the assumptions of Proposition 3.10. Furthermore assume that σ is independant of time and bounded along with its derivatives up to order 4. Then the market is ergodic.*

We start by proving the following lemma.

Lemma 3.12 (Construction of an intertwining operator). *There exists an elliptic operator such that*

$$\frac{\partial}{\partial\tau}\int \pi_\tau(S_0, dS)\,\phi(S) = \int \pi_\tau(S_0, dS)\,(\mathcal{R}_\tau\phi)(S) ; \tag{3.17}$$

$$0 < \alpha(S, \tau) < \exp(-2\delta\tau)\,|\sigma(S)|^2, \quad |\beta(\tau, s)| \le \exp(-\delta\tau)\,\sigma(S) . \tag{3.18}$$

Remark 3.13. If we take $\mathcal{R}_\tau = \mathcal{R}$ then (3.17) holds, but (3.18) fails.

Proof (of Lemma 3.12). By the semigroup property we have

$$\frac{\partial}{\partial\tau}\int \pi_\tau(S_0, dS)\,\phi(S) = \frac{1}{2}\sigma^2(S_0)\frac{d^2}{dS_0^2}\int \pi_\tau(S_0, dS)\,\phi(S) .$$

Using (3.16) and taking the conditional expectation with respect to $S_W(\tau)$ we get

$$\frac{1}{\sigma(S_0)} \frac{d}{dS_0} \int \pi_\tau(S_0, dS) \phi(S)$$

$$= \mathbb{E}^{S_W(0)=S_0} \left[\exp\left(\int_0^\tau \lambda(S_W(\xi)) d\xi \right) \sigma(S_W(\tau)) (\phi')(S_W(t)) \right]$$

$$= \int \pi_\tau(S_0, dS) \tilde{\beta}(S) \phi'(S) \quad \text{where } |\tilde{\beta}(S)| \leq c \exp(-\delta\tau)\sigma(S) .$$

Iterating this procedure we get an estimate of the second derivative which proves Lemma 3.12. □

Proof (of Theorem 3.11). Consider the following SDE with time-depending coefficients:

$$d\tilde{S}_W = \sqrt{\alpha(\tau, \tilde{S}_W(\tau))} \, dW(\tau) + \beta(\tau, \tilde{S}_W(\tau)) \, d\tau, \quad \tilde{S}_W(0) = S_0 . \qquad (3.19)$$

According to (3.17) the law of $\tilde{S}(\tau)$ is $\pi_\tau(S_0, \cdot)$. Furthermore by means of (3.18), for any $\tau, R > 0$,

$$\mathbb{E} \left| \int_{\tilde{S}_W(\tau)}^{\tilde{S}_W(\tau+R)} \frac{d\xi}{\sigma(\tilde{S}_W(\xi))} \right| \leq c \exp\left(-\frac{\delta}{2} R \right) ,$$

which implies first that $\lim_{\tau \to \infty} \tilde{S}_W(\tau)$ exists in terms of the distance χ on $\mathbb{R}$ defined by

$$\chi(S_1, S_2) := \int_{S_1}^{S_2} \frac{d\xi}{\sigma(\xi)} .$$

Since σ is bounded, convergence in the χ metric implies convergence in the usual metric of $\mathbb{R}$ as well. Therefore, we may conclude that

$$\rho_{S_0} := \lim_{\tau \to \infty} \pi_\tau(S_0, \cdot) \quad \text{exists.}$$

The fact that the market is without remote memory implies that $\rho_{S_0} = \rho_{S_1}$. See [57] for more details; [57] treats also the case $\mu \neq 0$. □

4

Multivariate Conditioning
and Regularity of Law

Given two random variables f and g on an abstract probability space, the theorem of Radon–Nikodym ensures existence of the conditional expectation $\mathbb{E}[f \mid g = a]$, almost everywhere in a. For Borel measures on a topological space, it is a basic problem to construct a *continuous version* of the function $a \mapsto \mathbb{E}[f \mid g = a]$. On the Wiener space, *quasi-sure analysis* [4, 144, 197] constructs, for g *non-degenerate*, continuous versions of conditional expectations.

This theoretical issue has a numerical counterpart in the question of how to compute a conditional expectation by Monte-Carlo simulation. The crude way of rejecting simulations giving trajectories which do not satisfy the conditioning turns out to be extremely costly, as far as computing time is concerned. The papers [78, 79], a preprint of Lions–Régnier, followed by applications to American options in [14, 42, 113, 181, 182], changed dramatically the Monte-Carlo computation of conditional expectations.

It is clear that existence of continuous versions for conditional expectations is very much linked to regularity of laws; such questions have been the first objective of the Stochastic Calculus of Variations [140, 141]. We emphasize in this chapter the case where the conditioning g is multivariate; the univariate case is treated at the end of the chapter as a special case of our multivariate study. In Sect. 4.5 the Riesz transform is introduced; it seems to be a new tool in this field.

4.1 Non-Degenerate Maps

We consider maps $g \colon \mathscr{W}^n \to \mathbb{R}^d$ with components $g^1, \dots, g^d$ such that each $g^i \in D_1^p(\mathscr{W}^n)$ for all $p < \infty$. The dimension d of the target space is called the *rank of g*. The *Malliavin covariance matrix* of g is the symmetric positive matrix defined by

$$\sigma_{ij}(W) := \sum_{k=1}^{n} \int_0^1 (D_{t,k}g^i)(W)\,(D_{t,k}g^j)(W)\,dt . \tag{4.1}$$

Definition 4.1. *A map* $g\colon \mathscr{W}^n \to \mathbb{R}^d$ *is said to be non-degenerate if*

$$\mathbb{E}\left[\det(\sigma)^{-p}\right] < \infty, \quad \forall\, p < \infty. \tag{4.2}$$

Definition 4.2 (Lifting up functions). *Given* $\phi \in C_b^1(\mathbb{R}^d)$, *the lift of* ϕ *to* $\mathscr{W}^n$ *is the function* $\tilde{\phi}$ *on* $\mathscr{W}^n$ *defined by* $\tilde{\phi} = \phi \circ g$. *Then*

$$\tilde{\phi} \in D_1^p(\mathscr{W}^n), \quad \forall\, p < \infty. \tag{4.3}$$

The operator $\phi \mapsto \tilde{\phi}$ *is denoted by* g^*.

Definition 4.3 (Pushing down functions). *Let* ν *be the law of* g, *that is* $\nu = g_*(\gamma)$ *is the direct image by* g *of the Wiener measure on* $\mathscr{W}^n$. *Then conditional expectation gives a map*

$$\mathbb{E}^g\colon L^p(\mathscr{W}^n; \gamma) \mapsto L^p(\mathbb{R}^d; \nu).$$

We call $\mathbb{E}^g[F]$ *the push-down of a function* $F \in L^p(\mathscr{W}^n; \gamma)$.

Pushing down is a left inverse of lifting up, i.e., $\mathbb{E}^g(\tilde{\phi}) = \phi$.

Definition 4.4 (Covering vector fields). *Let* z *be a vector field on* $\mathbb{R}^d$ *with components* $z^1, \ldots, z^d$. *An* $\mathbb{R}^d$-*valued process* $Z_W^1(t), \ldots, Z_W^d(t)$ *is called a covering vector field of* z *if*

$$\sum_{k=1}^n \int_0^1 Z_W^k(t)\,(D_{t,k}g^s)(W)\,dt = z_{g(W)}^s, \quad s = 1, \ldots, d. \tag{4.4}$$

Theorem 4.5. *Assume that* g *is non-degenerate. Then any vector field* z *on* $\mathbb{R}^d$ *has a unique covering vector field* $°Z$ *of minimal* L^2-*norm. Furthermore, the hypothesis*

$$z \in C_b^1(\mathbb{R}^d), \quad g \in D_2^p(\mathscr{W}^n), \quad \forall\, p < \infty,$$

implies that

$$°Z \in D_1^p(\mathscr{W}^n), \quad \forall\, p < \infty. \tag{4.5}$$

Proof. Denoting by (σ^{ik}) the inverse of the matrix (σ_{ik}), we define

$$°Z^k(t) = \sum_{0 \le s,\ell \le d} \sigma^{s\ell}\,(D_{t,k}g^s)\,z^\ell,$$

or symbolically,

$$°Z = \sum_{s,\ell} z^\ell \sigma^{\ell s} Dg^s. \tag{4.6}$$

Then we have

$$D_{\circ Z} g^q = \sum_k \int_0^1 Z^k(t)(D_{t,k} g^q) \, dt$$

$$= \sum_{0 \le \ell, s \le d} z^\ell \sigma^{s\ell} \sum_k \int_0^1 (D_{t,k} g^s)(D_{t,k} g^q) \, dt$$

$$= \sum_{s,\ell} \sigma_{qs} \sigma^{s\ell} z^\ell = z^q .$$

It remains to show that $^\circ Z$ defined by (4.6) provides the minimal norm. Consider another covering vector of the form $^\circ Z + Y$ where $D_Y g^\cdot = 0$. As $^\circ Z$ is a linear combination of Dg^s, the vanishing of $D_Y g^\cdot$ implies that $(Y|Dg^s) = 0$ for all s; thus $\|Y + {^\circ Z}\|^2 = \|Y\|^2 + \|{^\circ Z}\|^2 \ge \|{^\circ Z}\|^2$ as claimed. $\square$

In the sequel we short-hand computations on matrix indices; these computations are resumed in full details in the proof of (4.10) below. We get

$$D_{\tau,j} {^\circ Z} = -\sigma^{\cdots} z^\cdot (D_{\tau,j} \sigma_{\cdot,\cdot}) \sigma^{\cdots} z^\cdot (Dg^\cdot)$$

$$+ \sigma_{\cdot,\cdot} D_{\cdot,j}(Dg^\cdot) + \sigma^{\cdots} \frac{\partial z^\cdot}{\partial \xi^i}(D_{\tau,j} g^i);$$

$$D_{\tau,j} \sigma_{s\ell} = D_{\tau,j}(Dg^s | Dg^\ell)$$

$$= \sum_k \int_0^1 \left((D^2_{\tau,t;j,k} g^s) D_{t,k} g^\ell + (D^2_{\tau,t;j,k} g^\ell) D_{t,k} g^s \right) dt .$$

4.2 Divergences

Definition 4.6. *We say that a vector field z on $\mathbb{R}^d$ has a divergence $\vartheta_\nu(z)$ with respect to a probability measure ν, if the following formula of integration by parts holds:*

$$\int_{\mathbb{R}^d} \langle z, d\phi \rangle \, d\nu = \int_{\mathbb{R}^d} \phi \, \vartheta_\nu(z) \, d\nu, \quad \forall \phi \in C^1_b(\mathbb{R}^d), \tag{4.7}$$

where $\int_{\mathbb{R}^d} |\vartheta_\nu(z)| \, d\nu < \infty$.

Remark 4.7. Density of $C^1_b(\mathbb{R}^d)$ in the continuous functions of compact support implies uniqueness of the divergence.

Theorem 4.8 (Functoriality of the divergence operator). *Let g be a non-degenerate map, ν be the law of g, and let γ be the Wiener measure. For any covering vector field Z of z such that $\vartheta_\gamma(Z)$ exists, we have*

$$\vartheta_\nu(z) = \mathbb{E}^g[\vartheta_\gamma(Z)] . \tag{4.8}$$

Assuming furthermore that $g \in D_2^p(\mathcal{W}^n)$ for each $p < \infty$, then for any coordinate vector field $\partial/\partial\xi^i$,

$$\exists \vartheta_\nu \left(\frac{\partial}{\partial\xi}\right) \in L^p(\mathbb{R}^d; \nu), \quad \forall p < \infty . \tag{4.9}$$

Denoting by $\mathcal{N}$ the number operator defined in Theorem 1.33, then

$$\vartheta_\gamma(^\circ Z_i) = \sum_j \sigma^{ij} \mathcal{N}(g^j) + \sum_\ell (D^2 g^\ell)(^\circ Z_i, {}^\circ Z_\ell)$$

$$+ \sum_{k,\ell} \sigma^{ik}(D^2 g^k)(^\circ Z_\ell, Dg^\ell) . \tag{4.10}$$

Proof. We have the intertwining relation

$$D_Z(\tilde\phi) = \tilde u \quad \text{where } u = \langle z, d\phi\rangle =: \partial_z\phi ,$$

which symbolically may be written as

$$g^*(\partial_z\phi) = D_Z(g^*\phi) . \tag{4.11}$$

Note that this intertwining relation gives for the derivative of a composition of functions:

$$D_Z(\phi \circ g)(W) = \langle d\phi, g'(W)(Z)\rangle_{g(W)} = \langle d\phi, z\rangle_{g(W)} ,$$

and therefore

$$\int_{\mathbb{R}^d} \langle z, d\phi\rangle_\xi \, \nu(d\xi) = \mathbb{E}[\langle z, d\phi\rangle_{g(W)}]$$

$$= \mathbb{E}[D_Z\tilde\phi] = \mathbb{E}[\vartheta_\gamma(Z) \, \tilde\phi]$$

$$= \int_{\mathbb{R}^d} \phi(\xi) \, \mathbb{E}^{g=\xi}[\vartheta_\gamma(Z)] \, \nu(d\xi) .$$

We associate to the coordinate vector field $\partial/\partial\xi^i$ the minimal covering vector field $^\circ Z_i$; then we have $^\circ Z_i \in D_1^p(\mathcal{W}^n)$ by (4.5). Hence by means of (1.36) we conclude that $\vartheta_\gamma(^\circ Z_i)$ exists and belongs to L^p for any $p < \infty$. Formula (4.10) leads to an alternative proof of (4.9) by means of (1.51).

Thus it remains to prove (4.10). Note that $^\circ Z_i = \sum_j \sigma^{ij} Dg^j$ implies by (1.35) that

$$\vartheta(^\circ Z_j) = -\sum_j \sigma^{ij} \mathcal{N}(g^j) - \sum_j D_{Dg^j}(\sigma^{ij}) ,$$

where the first term of the r.h.s. comes from the identity $\mathcal{N}(\phi) = \vartheta(D\phi)$ (see Theorem 1.33). On the other hand, we have

$$-\sum_j (D_{Dg^j}(\sigma^{ij})) = \sum_{j,k,\ell} \sigma^{ik}(D_{Dg^j}\sigma_{k\ell})\sigma^{\ell j} ,$$

where

$$
\begin{aligned}
D_{Dg^j}\sigma_{k\ell} &= D_{Dg^j}(Dg^k|Dg^\ell) \\
&= (D^2g^k)(Dg^j, Dg^\ell) + (D^2g^\ell)(Dg^j, Dg^k) \\
&= \sum_{j,k,\ell} \sigma^{ik}\left[(D^2g^k)(Dg^j, Dg^\ell) + (D^2g^\ell)(Dg^j, Dg^k)\right]\sigma^{\ell j} \\
&= \sum_\ell (D^2g^\ell)(^\circ Z_i, {}^\circ Z_\ell) + \sum_{k,\ell} \sigma^{ik}(D^2g^k)(^\circ Z_\ell, Dg^\ell)
\end{aligned}
$$

which completes the proof.   □

Important comment. As the minimal covering vector field has a conceptual definition, one may wonder why we bothered ourselves with the general concept of covering vector fields. The reason is quite simple: the minimal covering vector field is in general not adapted to the Itô filtration; therefore a computation of its divergence requires the difficult task of computing a Skorokhod integral in an efficient way. In many cases it is however possible to find a predictable covering vector field with its divergence given by an Itô integral, which is easy to implement in a Monte-Carlo simulation.

4.3 Regularity of the Law of a Non-Degenerate Map

Let $\mathscr{S}(\mathbb{R}^d)$ be the Schwartz space of functions decreasing at infinity, along with all their derivatives, faster than any negative power of the Euclidean norm on $\mathbb{R}^d$. We further adopt the following notation:

$$
D^\infty(\mathscr{W}^n) := \bigcap_{(r,p)\in\mathbb{N}^2} D_r^p(\mathscr{W}^n). \tag{4.12}
$$

Theorem 4.9. *Let g be a non-degenerate map, and assume that*

$$
g^i \in D^\infty(\mathscr{W}^n), \quad \forall i \in \{1,\ldots,d\}. \tag{4.13}
$$

Then the law of g has a density with respect to the Lebesgue measure, which is infinitely differentiable and belongs to the Schwartz space $\mathscr{S}(\mathbb{R}^d)$.

Proof. Let **s** be a multi-index of length d of positive integers; denote by $|\mathbf{s}|$ the sum of the components. We associate to **s** the differential operator with constant coefficients defined by

$$
\partial_\mathbf{s} = \prod_{i=1}^d \left[\frac{\partial}{\partial\xi^i}\right]^{\mathbf{s}(i)}.
$$

Before continuing with the proof of Theorem 4.9 we formulate the following lemma.

Lemma 4.10. *For any multi-indices* **s***, there exists* $Q_{\mathbf{s}} \in D^{\infty}(\mathscr{W}^n)$ *such that the following formula of integration by parts holds for all* $\phi \in C_{\mathrm{b}}^{|\mathbf{s}|}(\mathbb{R}^d)$:

$$\mathbb{E}[g^*(\partial_{\mathbf{s}}\phi)] = \mathbb{E}[Q_{\mathbf{s}}\, g^*(\phi)]\,. \tag{4.14}$$

Proof. We proceed by induction on $|\mathbf{s}|$. For $|\mathbf{s}| = 1$ we may take $Q_i = \vartheta_\gamma(^{\circ}Z_i)$; it results from (4.10) that $Q_i \in D^{\infty}$.

Assume now that the lemma holds true for $|\mathbf{s}| < r$. Given **s** of length r, we can write

$$\partial_{\mathbf{s}} = \partial_{\mathbf{q}}\frac{\partial}{\partial \xi^i}, \quad |\mathbf{q}| = r - 1\,.$$

Defining $\phi_1 := \dfrac{\partial}{\partial \xi^i}\phi$, we get

$$\mathbb{E}[g^*(\partial_{\mathbf{s}}\phi)] = \mathbb{E}[g^*(\partial_{\mathbf{q}}\phi_1)] = \mathbb{E}[(g^*\phi)Q_{\mathbf{q}}] = \mathbb{E}[(D_{\circ Z_i}g^*\phi)\,Q_{\mathbf{q}}]$$
$$= \mathbb{E}[(g^*\phi)\,(\vartheta(^{\circ}Z_i)Q_{\mathbf{q}} - D_{\circ Z_i}(Q_{\mathbf{q}}))],$$

and therefore

$$Q_{\mathbf{s}} = \vartheta(^{\circ}Z_i)Q_{\mathbf{q}} - D_{\circ Z_i}(Q_{\mathbf{q}})\,, \tag{4.15}$$

which completes the proof. $\square$

Proof (of Theorem 4.9). Let u be the characteristic function of the law of g,

$$u(\eta) := \mathbb{E}[g^*\psi_\eta] \quad \text{where } \psi_\eta(\xi) = \exp(i\,(\xi|\eta)_{\mathbb{R}^d})\,.$$

Then, in terms of the Laplacian Δ on $\mathbb{R}^d$, we have

$$|\eta|_{\mathbb{R}^d}^{2m}u(\eta) = (-1)^m \mathbb{E}[g^*(\Delta^m\psi_\eta)]\,.$$

By means of (4.14), and taking into account that integration by parts of Δ^m is possible with the weight $Q_{\Delta^m} \in D^{\infty}$, we get

$$|\eta|_{\mathbb{R}^d}^{2m}u(\eta) = (-1)^m \mathbb{E}[(g^*\psi_\eta)\,Q_{\Delta^m}]$$

which gives

$$|u(\eta)| \leq \frac{\mathbb{E}[|Q_{\Delta^m}|]}{|\eta|^{2m}}\,. \tag{4.16}$$

Estimate (4.16) implies that the law of g has a density p belonging to $C_{\mathrm{b}}^{\infty}(\mathbb{R}^d)$.

As the Fourier transform preserves the space $\mathscr{S}(\mathbb{R}^d)$, it remains to show that $u \in \mathscr{S}(\mathbb{R}^d)$; to this end we must dominate all derivatives of u. The case of the first derivative is typical and may be treated as follows:

$$|\eta|^{2m}\frac{\partial u}{\partial \eta^1}(\eta) = (-1)^m i\, \mathbb{E}[g^*(\Delta^m\psi_\eta^1)],$$

where $\psi_\eta^1(\xi) = \xi^1 \exp(i\,(\xi|\eta))$. Integration by parts leads to the required domination:

$$\left|\frac{\partial u}{\partial \eta^1}(\eta)\right| \leq \frac{1}{|\xi|^{2m}}\mathbb{E}[|g^1|\,|Q_{\Delta^m}|] \leq \frac{1}{|\xi|^{2m}}\sqrt{\mathbb{E}[|g^1|^2]\mathbb{E}[|Q_{\Delta^m}|^2]}\,. \quad \square$$

4.4 Multivariate Conditioning

Theorem 4.11 (Pushing down smooth functionals). *Let g be an $\mathbb{R}^d$-valued non-degenerate map such that $g \in D^\infty(\mathscr{W}^n)$; let $p \in \mathscr{S}(\mathbb{R}^d)$ be the density of the law ν of g. To any $f \in D^\infty(\mathscr{W}^n)$, there exists $u_f \in \mathcal{S}(\mathbb{R}^d)$ such that the following disintegration formula holds:*

$$\mathbb{E}[f(g^*\phi)] = \int_{\mathbb{R}^d} \phi(\xi) u_f(\xi)\, d\xi, \quad \forall \phi \in C_b^0(\mathbb{R}^d) ; \tag{4.17}$$

furthermore

$$\mathbb{E}[f|\, g(W) = \xi] = \frac{u_f(\xi)}{p(\xi)} = \frac{u_f(\xi)}{u_1(\xi)}, \quad p(\xi) \neq 0 . \tag{4.18}$$

Remark 4.12. The conditional expectation in the l.h.s. of (4.18) is defined only almost surely, i.e., for $\xi \notin A$ where $\nu(A) = 0$. The r.h.s. however is a continuous function of ξ on the open set $p(\xi) \neq 0$; thus (4.18) provides indeed a *continuous version* of the conditional expectation.

Proof (of Theorem 4.11). We deal first with the special case

$$f \geq \varepsilon > 0, \quad \mathbb{E}[f] = 1 . \tag{4.19}$$

Let λ be the probability measure on $\mathscr{W}^n$, absolutely continuous with respect to the Wiener measure γ, which has f as its Radon–Nikodym derivative:

$$\frac{d\lambda}{d\gamma} = f . \tag{4.20}$$

Further denote by ρ the law of g under the measure λ, thus $\rho(A) = \mathbb{E}[f 1_A(g)]$ for any Borel set A of $\mathbb{R}^d$. Given a vector field Z on $\mathscr{W}^n$, its divergence $\vartheta_\lambda(Z)$ with respect to λ is defined by the formula of integration by parts

$$\int \Psi\, \vartheta_\lambda(Z)\, d\lambda = \int (D_Z \Psi)\, d\lambda .$$

Lemma 4.13. *Keeping the notation from above, we have*

$$\vartheta_\lambda(Z) = \vartheta_\gamma(Z) - D_Z \log f . \tag{4.21}$$

Proof (of Lemma 4.13). We may calculate as follows:

$$\int (D_Z \Psi)\, d\lambda = \int (D_Z \Psi)\, f\, d\gamma$$

$$= \int [D_Z(f\Psi) - f\Psi D_Z(\log f)]\, d\gamma$$

$$= \int [\vartheta_\gamma(Z)\, \Psi - \Psi D_Z(\log f)]\, f\, d\gamma. \quad \square$$

Proof (of Theorem 4.11). Let v be the characteristic function of ρ,

$$v(\eta) := \int_{\mathbb{R}^d} \exp(i(\eta|\xi))\, \rho d(\xi) = \int_{\mathscr{W}^n} (g^* \psi_\eta)\, d\lambda \ ;$$

then

$$\eta^1 v(\eta) = \int_{\mathscr{W}^n} D_{\circ Z_1}(g^* \psi_\eta)\, d\lambda = \int_{\mathscr{W}^n} \vartheta_\lambda(^\circ Z_1)(g^* \psi_\eta)\, d\lambda \ .$$

Using (4.21) we get

$$\eta^1 v(\eta) = \mathbb{E}\left[f(\vartheta_\gamma(^\circ Z_1) - D_{\circ Z_1} \log f)(g^* \psi_\eta) \right] \ ,$$

from which we conclude that $|v(\eta)| < c/|\eta^1|$. Iterating this procedure of integration by parts, we see as before that $v \in \mathscr{S}(\mathbb{R}^d)$.

It remains to remove assumption (4.19). Note that the condition $\mathbb{E}[f] = 1$ is easy to satisfy through multiplying f by a constant. We can find two functions $\chi_1, \chi_2 \in C_b^\infty(\mathbb{R})$ such that $\chi_1 + \chi_2 = 1$, $\mathrm{supp}(\chi_1) \subset [-1, \infty[$ and $\mathrm{supp}(\chi_2) \subset]-\infty, 1]$. For f given, we define

$$f_1 = f \cdot \chi_1(f) + 2, \quad f_2 = -f \cdot \chi_2(f) + 2 \ .$$

Then f_1, f_2 satisfy the inequalities in (4.19); hence there exist $u_{f_i} \in \mathscr{S}(\mathbb{R}^d)$ satisfying (4.17). As relation (4.17) is linear, it may be satisfied by taking

$$u_f = u_{f_1} - u_{f_2} \ .$$

The first equality in (4.18) is a consequence of the smoothing property of the conditional expectation; the second one results from applying the first equation to the function $f = 1$. $\square$

The following theorems provide more effective formulae for computing the relevant conditional expectations. Indeed, by means of (4.18) the computation of the conditional expectation may be reduced to the computation of $u_f(a)$.

We shall use the following assumptions:

$$g \colon \mathscr{W}^n \to \mathbb{R}^d \text{ is a non-degenerate map, } g \in D^\infty \text{ and } f \in D^\infty. \quad (4.22)$$

Let H be the Heaviside function defined by $H(t) = 1$ for $t > 0$ and $H(t) = 0$ otherwise. Finally set $^\alpha H(t) = H(t - \alpha)$.

Theorem 4.14. *Let $\mathcal{R}_g^i \colon D^\infty \to D^\infty$ be the first order differential operator defined by*

$$\mathcal{R}_g^i(\Psi) = \vartheta(Z_i)\Psi - D_{Z_i}\Psi$$

where Z_i are smooth covering vector fields of the coordinate vector field of $\mathbb{R}^d$. Under assumption (4.22) and assuming that $p(a) \neq 0$ we have

$$u_f(a) = \mathbb{E}\left[\Gamma_g(f)\, g^* \left(\prod {}^{a^i} H(\xi^i) \right) \right]; \quad \Gamma_g(f) = (\mathcal{R}_g^1 \circ \ldots \circ \mathcal{R}_g^d)(f) \ . \quad (4.23)$$

Remark 4.15. We emphasize the important fact that the weight $\Gamma_g(f)$ does not depend on the values of the conditioning a.

The proof of Theorem 4.14 will be based on an approximation scheme. Consider the convolution $H_\varepsilon := H * \psi_\varepsilon$ where ψ_ε is a mollifier: $\psi_\varepsilon(t) = \varepsilon^{-1}\psi(t/\varepsilon)$, ψ being a C^∞-function supported in $[-1, 0]$ with integral equal to 1. Denote by H'_ε the derivative of H_ε.

Lemma 4.16. *The function u_f defined by (4.17) has the expression:*

$$u_f(a) = \lim_{\varepsilon_1 \to 0} \cdots \lim_{\varepsilon_d \to 0} \mathbb{E}\left[f \prod_{i=1}^{d} {}^{a_i} H'_{\varepsilon_i}(g^i)\right]$$

$$= \lim_{\varepsilon_1 \to 0} \cdots \lim_{\varepsilon_d \to 0} \mathbb{E}\left[f\, D_{Z_1} \ldots D_{Z_d} g^* \left(\prod_{i=1}^{d} {}^{a_i} H_{\varepsilon_i}(\xi_i)\right)\right]. \tag{4.24}$$

Proof. We short-hand the notation by taking $a = 0$ and drop the indices a_i. Using (4.11) and (4.17), we have

$$\mathbb{E}\left[f \prod_{i=1}^{d} H'_{\varepsilon_i}(g^i)\right] = \int_{\mathbb{R}^d} \prod_{i=1}^{d} H'_{\varepsilon_i}(\xi^i)\, u_f(\xi)\, d\xi \, ,$$

and thus

$$\lim_{\varepsilon_1 \to 0} \cdots \lim_{\varepsilon_d \to 0} \int_{\mathbb{R}^d} \prod_{i=1}^{d} H'_{\varepsilon_i}(\xi^i)\, u_f(\xi)\, d\xi = u_f(0) \, .$$

On the other hand

$$\mathbb{E}\left[f\, D_{Z_1} \ldots D_{Z_d} g^* \left(\prod_{i=1}^{d} {}^{a_i} H_{\varepsilon_i}(\xi_i)\right)\right]$$

$$= \mathbb{E}\left[f\, D_{Z_1} \ldots D_{Z_{d-1}} g^* \left(\frac{\partial}{\partial \xi^d} \prod_{i=1}^{d} {}^{a_i} H_{\varepsilon_i}(\xi_i)\right)\right]$$

$$= \mathbb{E}\left[f\, g^* \left(\frac{\partial}{\partial \xi^1} \cdots \frac{\partial}{\partial \xi^d} \prod_{i=1}^{d} {}^{a_i} H_{\varepsilon_i}(\xi_i)\right)\right]. \quad \square$$

Recall that the rank of a non-degenerate map is the dimension d of the target space. We prove Theorem 4.14 by induction on d. The case $d = 1$ is covered by the following theorem.

Theorem 4.17. *For $d = 1$, under assumption (4.22), denote by Z_1 a covering vector field of the coordinate vector field. Assume that $\vartheta(Z_1)$ exists and that $p(0) \neq 0$. Then (4.23) holds true, which means that*

$$u_f(a) = \mathbb{E}\left[{}^a H(g)(f\vartheta(Z_1) - D_{Z_1} f)\right]. \tag{4.25}$$

Remark 4.18. We may take as covering vector field the minimal covering vector field

$$°Z_1 = \frac{Dg^1}{\|Dg^1\|^2} \,.$$

This will be developed in Sect. 4.6.

Proof (of Theorem 4.17). Starting from (4.23), we have

$$\mathbb{E}\left[f\, D_{Z_1} g^*(H_\varepsilon)\right] = \mathbb{E}\left[(f\, \vartheta(Z_1) - D_{Z_1} f)\, g^*(H_\varepsilon)\right] \,. \tag{4.26}$$

Thus letting $\varepsilon \to 0$, we find (4.25). $\square$

Proof (of Theorem 4.14 by induction). The induction is done on the validity of $(4.23)_r$ for a non-degenerate map g of rank $\leq r$.

Note that $(4.23)_1$ holds true as a consequence of (4.25). Assume now that $(4.28)_r$ hold true for all $r < d$. Denote by $(g^1, \ldots, g^d)$ a non-degenerate map of rank d, and let h be the $\mathbb{R}^{d-1}$-valued map $h = (g^2, \ldots, g^d)$.

Lemma 4.19. *The map h is non-degenerate.*

Proof (of Lemma 4.19). Denote by σ_g, σ_h the corresponding covariance matrices, then

$$\det(\sigma_h)\, \|Dg^1\|^2 \geq \det(\sigma_g) \,,$$

therefore

$$\|\det(\sigma_h)^{-1}\|_{L^p} \leq \|Dg^1\|_{L^{2p}}\, \|\det(\sigma_f)^{-1}\|_{L^{2p}}.   \square$$

Consider $g = (g^1, h)$. By induction hypothesis, there exists a differential operator Γ_h such that identity $(4.23)_1$ holds true. Let $\{Z_i : i = 1, \ldots, d\}$ be covering vector fields for the coordinate vector fields of g. We compute Γ_h by taking Z_i, $i = 2, \ldots, d$, as covering vector fields. Then by (4.24) we have

$$u_f(0) = \lim_{\varepsilon_1 \to 0}\left(\lim_{\varepsilon_2,\ldots,\varepsilon_d \to 0} \mathbb{E}\left[f\, H'_{\varepsilon_1}(g^1) \prod_{i=2}^{d} H'_{\varepsilon_i}(g^i)\right]\right) \,.$$

Denoting $\tilde{f} = f\, H'_{\varepsilon_1}(g^1)$, we get

$$\lim_{\varepsilon_2,\ldots,\varepsilon_d \to 0} \mathbb{E}\left[\tilde{f} \prod_{i=2}^{d} H'_{\varepsilon_i}(g^i)\right] = \lim_{\varepsilon_2,\ldots,\varepsilon_d \to 0} \mathbb{E}\left[\tilde{f} \prod_{i=2}^{d} H'_{\varepsilon_i}(h^i)\right]$$

$$= \mathbb{E}\left[\Gamma_h(\tilde{f}) \prod_{i=2}^{d} H(h^i)\right] \,,$$

and therefore

$$u_f(0) = \lim_{\varepsilon_1 \to 0} \mathbb{E}\left[\Gamma_h(f\, H'_{\varepsilon_1}(g^1)) \prod_{i=2}^{d} H(g^i)\right] \,. \tag{4.27}$$

Formula (4.27) will be further evaluated by means of the following lemma.

Lemma 4.20. *For any smooth function φ the following formula holds:*

$$\Gamma_h(\varphi(g^1)\,f) = \varphi(g^1)\Gamma_h(f)\,. \tag{4.28}$$

Proof (of Lemma 4.20). Using the product formula (4.23) it is sufficient to prove

$$\mathcal{R}_h^s(\varphi(g^1)\,f) = \varphi(g^1)\mathcal{R}_h^s(f), \quad s = 2,\dots,d\,.$$

This commutation is equivalent to $D_{Z_s}(g^1) = 0$ and holds true because the Z_i have been chosen to be a system of covering coordinate vector fields with respect to g.

Proof (of Theorem 4.14; conclusion). By means of Lemma 4.20, formula (4.27) can be written as

$$u_f(0) = \lim_{\varepsilon_1 \to 0} \mathbb{E}\left[H'_{\varepsilon_1}(g^1)\,\Gamma_h(f)\prod_{i=2}^{d} H(g^i)\right]$$

$$= \mathbb{E}\left[\mathcal{R}_{g^1}\left(\Gamma_h(f)\prod_{i=2}^{d} H(g^i)\right) H(g^1)\right]$$

where the last equality comes from an application of $(4.23)_1$ to g^1. As in (4.28) we have now

$$\mathcal{R}_{g^1}\left(\Gamma_h(f)\prod_{i=2}^{d} H(g^i)\right) = \left(\prod_{i=2}^{d} H(g^i)\right) \times \mathcal{R}_{g^1}\left(\Gamma_h(f)\right),$$

and hence,

$$u_f(0) = \mathbb{E}\left[\mathcal{R}_g^1(\Gamma_h(f))\prod_{i=1}^{d} H(g^i)\right]. \quad \square$$

Example 4.21. We compute the differential operator Γ_g in the bivariate case of (4.23):

$$\Gamma_g = (\vartheta(Z_1) - D_{Z_1})(\vartheta(Z_2) - D_{Z_2}) \tag{4.29}$$
$$= D_{Z_1}D_{Z_2} - [\vartheta(Z_1)D_{Z_2} + \vartheta(Z_2)D_{Z_1}] + [\vartheta(Z_1)\vartheta(Z_2) - (D_{Z_1}\vartheta(Z_2))].$$

For an explicit formula one has to calculate $D_{Z_1}\vartheta(Z_2)$. This has been done in Sect. 1.7.

4.5 Riesz Transform and Multivariate Conditioning

The drawback of the methodology used in Sect. 4.4 is that one has to compute iterative derivatives and derivatives of divergences. In this section we shall propose an alternative formula where the divergences no longer need to be differentiated. The price to pay is that the bounded Heaviside function H must be replaced by an unbounded kernel derived from the Newtonian potential; in a Monte-Carlo simulation thus trajectories ending near the required conditioning get a preferred weight, a fact which does not seem unreasonable.

Theorem 4.22 (Construction of the Riesz transform). *Consider on* $\mathbb{R}^d$ *the kernels*

$$\mathcal{C}_i(\xi) = -c_d \frac{\xi_i}{\|\xi\|^{d-1}}$$

where $\|\xi\|^2 = \sum_i \xi_i^2$ *and* $c_d = 2(d-2)/a(d)$ *for* $d > 2$, $c_2 = 2/a(2)$; *here* $a(d)$ *denotes the area of the unit sphere of* $\mathbb{R}^d$ *(i.e.* $a(2) = 2\pi$, $a(3) = 4\pi, \ldots$). *Furthermore, denote by* $*$ *the convolution of functions on* $\mathbb{R}^d$. *Then, for any* $h \in C^1(\mathbb{R}^d)$ *with compact support, we have*

$$h(\xi) = \sum_{i=1}^{d} \left(\mathcal{C}_i * \frac{\partial h}{\partial \xi_i} \right)(\xi), \quad \xi \in \mathbb{R}^d. \tag{4.30}$$

Proof. We give a proof that is valid under the more stringent hypothesis that $f \in C^2$ (see for instance [143] for a proof in the general case). Consider the Newton potential kernel $q_d(\xi)$ defined as

$$q_2(\xi) = \log \frac{1}{\|\xi\|} \quad \text{and} \quad q_d(\xi) = \|\xi\|^{2-d}, \quad \text{for } d > 2.$$

We remark that

$$\frac{\partial}{\partial \xi_k} q_d = a(d)\, \mathcal{C}_k, \quad q \geq 2.$$

The r.h.s of (4.30) can be written as

$$\frac{1}{a(d)} \sum_{k=1}^{d} \frac{\partial q_d}{\partial \xi_k} * \frac{\partial f}{\partial \xi_k} = \frac{1}{a(d)} q_d * \left(\sum_{k=1}^{d} \frac{\partial^2 f}{\partial \xi_k^2} \right) = \frac{1}{a(d)} q_d * \Delta f,$$

where we used the identity linking convolution and derivations $\frac{\partial u}{\partial \xi_k} * v = u * \frac{\partial v}{\partial \xi_k}$. The conclusion follows now from the fact that $q_d/a(d)$ is the fundamental solution of the Laplace equation. $\square$

Keeping the notations of Sect. 4.4, we can state the following theorem on conditioning.

Theorem 4.23. *Let* $\mathcal{R}_g^i \colon D^\infty \to D^\infty$ *be the first order differential operator defined by*

$$\mathcal{R}_g^i(\Psi) = \vartheta(Z_i)\Psi - D_{Z_i}\Psi$$

where Z_i *are smooth covering vector fields of the coordinate vector field of* $\mathbb{R}^d$. *Assuming that* $p(a) \neq 0$ *we have*

$$u_f(a) = c_d \sum_i \mathbb{E}\left[\mathcal{R}_g^i(f) \frac{g^i - a^i}{\|g - a\|^{d-2}} \right]. \tag{4.31}$$

Proof. It is sufficient, as seen before, to prove the theorem for $f > 0$. Applying the preceeding theorem with u_f for h, we get

$$u_f(a) = c_d \sum_i \int_{\mathbb{R}^d} \frac{\partial}{\partial \xi_i} u_f(\xi) \, \frac{\xi^i - a^i}{\|\xi - a\|^{d-1}} \, d\xi$$

$$= c_d \sum_i \int_{\mathbb{R}^d} \frac{\partial}{\partial \xi_i} \log u_f(\xi) \, \frac{\xi^i - a^i}{\|\xi - a\|^{d-1}} \, u_f(\xi) \, d\xi.$$

Thus, in terms of the image measure $\nu(d\xi) = u_f(\xi) \, d\xi$, we get

$$u_f(a) = c_d \sum_i \int_{\mathbb{R}^d} \vartheta_\nu \left(\frac{\partial}{\partial \xi_i} \right) \frac{\xi^i - a^i}{\|\xi - a\|^{d-1}} \, \nu(d\xi)$$

$$= c_d \sum_i \mathbb{E} \left[f \, \vartheta_{f\mu}(Z_i) \, \frac{g^i - a^i}{\|g - a\|^{d-2}} \right],$$

where the fact that $\nu = g_*(f\mu)$ and the functoriality principle for divergences have been used for the last equality. The proof is completed by the following identity, written for an arbitrary test function v:

$$\mathbb{E}[f \, \vartheta_{f\mu}(Z_i) \, v] = \mathbb{E}[f D_Z v] = \mathbb{E}[D_Z(fv) - v D_Z f] = \mathbb{E}[v \mathcal{R}(f)]. \quad \square$$

4.6 Example of the Univariate Conditioning

We could specialize the theorems obtained in the previous sections to the one-dimensional case $d = 1$, except for the results of Sect. 4.5 which are valid only for $d \geq 2$. We prefer instead to give new proofs starting from scratch, hoping that this more elementary perspective sheds additional light on the results of the previous sections.

Theorem 4.24. *Let ϕ be a real-valued random variable such that $\phi \in D_2^p(\mathscr{W})$. Denote $\|D\phi\|^2 = \int_0^1 |D_t\phi|^2 \, dt$ and assume that there exists $\varepsilon > 0$ such that*

$$\mathbb{E} \left[\|D\phi\|^{-(2q+\varepsilon)} \right] < \infty, \quad 1/p + 1/q = 1 \,. \tag{4.32}$$

Then the law $P \circ \phi^{-1}$ of ϕ has a continuous density u with respect to the Lebesgue measure.

Proof. Consider the vector field Z on $\mathscr{W}$ defined by $Z = D\phi/\|D\phi\|^2$. Then we have

$$(Z \,|\, D\phi) = 1 \,. \tag{4.33}$$

Now assume that the following condition for the divergence $\vartheta(Z)$ of Z holds true:

$$\mathbb{E}[|\vartheta(Z)|] < \infty \,. \tag{4.34}$$

Considering the sequence of continuous functions

$$u_n^{\xi_0}(\xi) = \begin{cases} 0, & \text{for } \xi \le \xi_0 - \frac{1}{n} , \\ 1, & \text{for } \xi \ge \xi_0 + \frac{1}{n} , \\ \text{linear}, & \text{between } \xi_0 - \frac{1}{n} \text{ and } \xi_0 + \frac{1}{n}, \end{cases}$$

and denoting by ν the law of ϕ, we have

$$\frac{2}{n} \int_{\xi_0 - 1/n}^{\xi_0 + 1/n} \nu(d\xi) = \mathbb{E}\left[(u_n^{\xi_0})'(\phi)\right] = \mathbb{E}\left[D_Z(u_n^{\xi_0} \circ \phi)\right] = \mathbb{E}\left[(u_n^{\xi_0} \circ \phi)\,\vartheta(Z)\right] .$$

Letting $n \to \infty$, we get that ν has a density u and that

$$u(\xi_0) = \mathbb{E}\left[\vartheta(Z)1_{\{\phi(W) > \xi_0\}}\right] .$$

The Lebesgue theorem of monotone convergence then implies that

$$\lim_{\xi \to \xi_0,\, \xi > \xi_0} u(\xi) = u(\xi_0), \qquad \lim_{\xi \to \xi_0,\, \xi < \xi_0} p(\xi) = u(\xi_0) .$$

It remains to prove (4.34). To this end we note that

$$(D_\tau Z)(t) = -\left(2\,\frac{D_\tau \phi}{\|D\phi\|^3}\right) D_t(\phi) + \frac{1}{\|D\phi\|^2}\, D_{\tau,t}^2 \phi ;$$

$$\int_0^1 \int_0^1 |D_\tau Z(t)|^2 \, d\tau dt \le 2(A + B) ,$$

where

$$A = \frac{4}{\|D\phi\|^6} \left(\int_0^1 \|D_\tau \phi\|^2 \, d\tau\right)^2 = \frac{4}{\|D\phi\|^2(W)} ,$$

$$B = \frac{1}{\|D\phi\|^4(W)} \int_0^1 \int_0^1 |D_{\tau,t}^2 \phi|^2 \, d\tau dt = \frac{\|D^2\phi\|^2(W)}{\|D\phi\|^4(W)} .$$

Using Hölder's inequality

$$\mathbb{E}\left[\sqrt{B}\right] \le \|\phi\|_{D_2^p}\, \mathbb{E}\left[\|D\phi\|^{-2q}\right]^{1/q} ,$$

as $\varepsilon > 0$ in the hypothesis, we deduce the existence of $\eta > 0$ such that $\mathbb{E}[B^{(1+\eta)/2}] < \infty$. Since

$$\|Z\|_{D_1^{1+\eta}} \le \mathbb{E}[(A + B)^{(1+\eta)/2}] ,$$

we get $\|Z\|_{D_1^{1+\eta}} < \infty$. Along with (1.37), $\mathbb{E}\left[|\vartheta(Z)|^{1+\eta}\right] < \infty$ is obtained. $\square$

Corollary 4.25. *The decay of u at infinity is dominated by*

$$\|\vartheta(Z)\|_{L^{p'}}\, \gamma(\{\phi > a\})^{1/q'}, \quad 1/p' + 1/q' = 1 . \tag{4.35}$$

Theorem 4.26. *Let $\Psi \in D_1^q(\mathscr{W})$. The conditional expectation $\mathbb{E}[\Psi|\phi = a]$ is a continuous function of a; in terms of the notation of the previous theorem it is given by*

$$\mathbb{E}[\Psi|\phi = a] = \frac{1}{u(a)} \mathbb{E}\left[(\Psi\vartheta(Z) - D_Z\Psi)1_{\{\phi > a\}}\right] . \tag{4.36}$$

Proof. We keep the notation of Theorem 4.24. Then we have

$$u(a)\,\mathbb{E}[\Psi|\phi = a] = \lim_{n\to\infty} \mathbb{E}\left[\Psi\,(u_n^{\xi_0})'(\phi)\right] ,$$

where

$$\mathbb{E}\left[\Psi\,(u_n^{\xi_0})'(\phi)\right] = \mathbb{E}[\Psi D_Z(u_n^{\xi_0} \circ \phi)]$$
$$= \mathbb{E}\left[\Psi(u_n^{\xi_0} \circ \phi)\,\vartheta(Z)\right] - \mathbb{E}\left[(u_n^{\xi_0} \circ \phi)D_Z\Psi\right] .$$

This proves the claim. $\quad\square$

Theorem 4.20. (cf. [OW]). The combinatorial equation $SP^\infty(X) \simeq$ ...

$$\Phi^G_*(\ldots) = \frac{1}{|G|} \sum_{g} \ldots$$

5

Non-Elliptic Markets and Instability in HJM Models

In this chapter we drop the ellipticity assumption which served as a basic hypothesis in Chap. 3 and in Chap. 2, except in Sect. 2.2.

We give up ellipticity in order to be able to deal with models with random interest rates driven by Brownian motion (see [61] and [104]). The empirical market of interest rates satisfies the following two facts which rule out the ellipticity paradigm:

1) high dimensionality of the state space constituted by the values of bonds at a large numbers of distinct maturities;
2) low dimensionality variance which, by empirical variance analysis, within experimental error of 98/100, leads to not more than 4 independent scalar-valued Brownian motions, describing the noise driving this high-dimensional system (see [41]).

Elliptic models are therefore ruled out and hypoelliptic models are then the most regular models still available. We shall show that these models display structural instability in smearing instantaneous derivatives which implies an unstable hedging of digital options.

Practitioners hedging a contingent claim on a single asset try to use all trading opportunities inside the market. In interest rate models practitioners will be reluctant to hedge a contingent claim written under bounds having a maturity less than five years by trading contingent claims written under bounds of maturity 20 years and more. This quite different behaviour has been pointed out by R. Cont [52] and R. Carmona [48].

R. Carmona and M. Tehranchi [49] have shown that this empirical fact can be explained through models driven by an infinite number of Brownian motions. We shall propose in Sect. 5.6 another explanation based on the progressive smoothing effect of the heat semigroup associated to a hypoelliptic operator, an effect which we call *compartmentation*.

This infinite dimensionality phenomena is at the root of modelling the interest curve process: indeed it has been shown in [72] that the interest rate model process has very few finite-dimensional realizations.

Section 5.7 develops for the interest rate curve a method similar to the methodology of the price-volatility feedback rate (see Chap. 3). We start by stating the possibility of measuring in real time, in a highly traded market, the full historical volatility matrix: indeed cross-volatility between the prices of bonds at two different maturities has an economic meaning (see [93, 94]). As the market is highly non-elliptic, the multivariate price-volatility feedback rate constructed in [19] cannot be used. We substitute a pathwise econometric computation of the bracket of the driving vector of the diffusion. The question of efficiency of these mathematical objects to decipher the state of the market requires numerical simulation on intra-day ephemerides leading to stable results at a properly chosen time scale.

5.1 Notation for Diffusions on $\mathbb{R}^N$

We start by recalling the notation of Sect. 2.2. On the space $\mathbb{R}^N$ (N will usually be large) the coordinates of points $r \in \mathbb{R}^N$ are denoted r^ξ, $\xi = 1, \ldots, N$. Given $(n+1)$ smooth vector fields $A_0, \ldots, A_n$ on $\mathbb{R}^N$ (the case $n << N$ being not excluded), the ξth component of A_k is denoted A_k^ξ.

We shall work on $\mathbb{R}^N$, but using infinite-dimensional Hilbert space techniques; in practice N will be finite but large. This implies that mathematically well-defined objects, as the determinant of an $N \times N$ matrix, may become numerically unstable. Having this "infinite-dimensional" point of view in mind, we resume some computations already done in Chap. 5.

Given n scalar-valued Brownian motions $W_1, \ldots, W_n$ we shall deal with the following SDE:

$$dr_W(t) = \sum_{k=1}^{n} A_k(r_W(t)) \, dW_k(t) + A_0(r_W(t)) \, dt \ . \tag{5.1}$$

Resuming in this situation the computations of Sect. 2.2, we associate to a vector field A_k the function $\mathbf{A}_k \colon \mathbb{R}^N \mapsto \mathcal{M}_N$ taking values in the space of real $N \times N$ matrices, defined as

$$(\mathbf{A}_k)_\eta^\xi := \frac{\partial A_k^\xi}{\partial r^\eta} \ .$$

For $t > t_0$, let $U_{t \leftarrow t_0}^W(r_0)$ be the solution of (5.1) with initial value r_0 at time t_0. The linearized equation, again denoted

$$\zeta_W(t) = J_{t \leftarrow t_0}^W(\zeta_0) \ ,$$

is given by

$$d_t(\zeta_W(t)) = \sum_{k=1}^{n} \mathbf{A}_k(\zeta_W(t)) \, dW_K(t) + \mathbf{A}_0(\zeta_W(t)) \, dt \ . \tag{5.2}$$

Theorem 5.1. *The map $r \mapsto U_{t \leftarrow t_0}^W(r)$ is a flow of diffeomorphisms satisfying the composition rule*

$$U_{t \leftarrow t_1}^W \circ U_{t_1 \leftarrow t_0}^W = U_{t \leftarrow t_0}^W, \quad t_0 \leq t_1 \leq t \,;$$

we define $U_{t \leftarrow t_0}^W$ as the inverse map of $U_{t_0 \leftarrow t}^W$.

Proof. See Kunita [116], Malliavin [144], Nualart [159]. □

5.2 The Malliavin Covariance Matrix of a Hypoelliptic Diffusion

Fix $W \in \mathscr{W}^n$, the space of continuous paths from $[0,T]$ to $\mathbb{R}^n$ vanishing at 0. The *backward smearing operator* is by definition the map $\mathcal{Q}_W^{\leftarrow}$ which assigns to each $\theta \in L^2([t_0, T]; \mathbb{R}^n)$ the vector $\mathcal{Q}_W^{\leftarrow}(\theta)$ in $\mathbb{R}^N$ defined by

$$\mathcal{Q}_W^{\leftarrow}(\theta) = \int_0^T \sum_{k=1}^n J_{t_0 \leftarrow t}^W A_k(r_W(t)) \, \theta^k(t) \, dt \,; \tag{5.3}$$

analogously, the *forward smearing operator* $\mathcal{Q}_W^{\rightarrow}$ is defined by

$$\mathcal{Q}_W^{\rightarrow}(\theta) = \int_0^T \sum_{k=1}^n J_{T \leftarrow t}^W A_k(r_W(t)) \, \theta^k(t) \, dt \,. \tag{5.4}$$

As $\mathbb{R}^N$ and $L^2([t_0, T]; \mathbb{R}^n)$ carry natural Hilbert space structures, the adjoint $(\mathcal{Q}_W^{\rightarrow})^*$ is a well-defined object.

The *forward Malliavin covariance matrix* is defined as the symmetric $N \times N$ matrix $(\sigma_W^{\rightarrow})^{\xi,\eta}$ associated to the hermitian operator $\sigma_W^{\rightarrow} := \mathcal{Q}_W^{\rightarrow} \circ (\mathcal{Q}_W^{\rightarrow})^*$ on $\mathbb{R}^N$:

$$(\sigma_W^{\rightarrow})^{\xi,\eta} = \int_{t_0}^T \sum_{k=1}^n \big(e^\xi \,\big|\, J_{T \leftarrow t}^W A_k(r_W(t))\big)_{\mathbb{R}^N} \big(e^\eta \,\big|\, J_{T \leftarrow t}^W A_k(r_W(t))\big)_{\mathbb{R}^N} dt \tag{5.5}$$

where $\{e^\xi\}$ denotes the canonical basis of $\mathbb{R}^N$. The associated quadratic form reads as

$$\big(\sigma_W^{\rightarrow}(\zeta) \,\big|\, \zeta\big)_{\mathbb{R}^N} = \int_{t_0}^T \sum_{k=1}^n \big(\zeta \,\big|\, J_{T \leftarrow t}^W A_k(r_W(t))\big)_{\mathbb{R}^N}^2 \, dt \,. \tag{5.6}$$

The *backward Malliavin covariance matrix* is the symmetric $N \times N$ matrix $(\sigma_W^{\leftarrow})^{\xi,\eta}$ associated to the hermitian operator on $\mathbb{R}^N$ defined by

$$\sigma_W^{\leftarrow} := \mathcal{Q}_W^{\leftarrow} \circ (\mathcal{Q}_W^{\leftarrow})^* \,. \tag{5.7}$$

The backward covariance matrix is given by the following formula:

$$(\sigma_{\overrightarrow{W}})^{\xi,\eta} = \int_{t_0}^{T} \sum_{k=1}^{n} (e^{\xi} \,|\, J_{t_0 \leftarrow t}^{W} A_k(r_W(t)))_{\mathbb{R}^N} (e^{\eta} \,|\, J_{t_0 \leftarrow t}^{W} A_k(r_W(t)))_{\mathbb{R}^N} \, dt$$

$$= \int_{t_0}^{T} \sum_{k=1}^{n} ((J_{t_0 \leftarrow t}^{W})^{*}(e^{z}\xi) \,|\, A_k(r_W(t)))_{\mathbb{R}^N} ((J_{t_0 \leftarrow t}^{W})^{*}(e^{\eta}) \,|\, A_k(r_W(t)))_{\mathbb{R}^N} \, dt.$$

$$(5.8)$$

The relation $J_{T \leftarrow t_0}^{W} \circ J_{t_0 \leftarrow t}^{W} = J_{T \leftarrow t}^{W}$ combined with (5.7), (5.8) implies the following conjugation between the forward and backward covariance matrices:

$$(J_{T \leftarrow t_0}^{W})^{*} \circ \sigma_{\overrightarrow{W}} \circ (J_{T \leftarrow t_0}^{W})^{*} = \sigma_{\overrightarrow{W}} . \tag{5.9}$$

As the matrix $J_{T \leftarrow t_0}^{W}$ is invertible, invertibility of $\sigma_{\overrightarrow{W}}$ is equivalent to the invertibility of $\sigma_{\overleftarrow{W}}$.

Theorem 5.2. *Let the forward covariance matrix $\sigma_{\overrightarrow{W}}$ be invertible.*

i) *For any $\zeta \in \mathbb{R}^N$, defining $\theta_{\overrightarrow{W}} = ((Q^{\rightarrow})^{*} \circ (\sigma_{\overrightarrow{W}})^{-1})(\zeta)$, we have*

$$Q_{\overrightarrow{W}}(\theta_{\overrightarrow{W}}) = \zeta; \tag{5.10a}$$

if $\lambda_{\overrightarrow{W}}$ denotes the smallest eigenvalue of $\sigma_{\overrightarrow{W}}$ then

$$\|\theta_{\overrightarrow{W}}\|_{L^2}^2 = ((\sigma_{\overrightarrow{W}})^{-1}\zeta \,|\, \zeta) \leq (\lambda_{\overrightarrow{W}})^{-1}\|\zeta\|^2 . \tag{5.10b}$$

ii) *For any $\zeta \in \mathbb{R}^N$, defining $\theta_{\overleftarrow{W}} = ((Q^{\leftarrow})^{*} \circ (\sigma_{\overleftarrow{W}})^{-1})(\zeta)$, we have*

$$Q_{\overleftarrow{W}}(\theta_{\overleftarrow{W}}) = \zeta ; \tag{5.11a}$$

if $\lambda_{\overleftarrow{W}}$ denotes the smallest eigenvalue of $\sigma_{\overleftarrow{W}}$ then

$$\|\theta_{\zeta}^{\leftarrow}\|_{L^2}^2 = ((\sigma_{\overleftarrow{W}})^{-1}\zeta \,|\, \zeta) \leq (\lambda_{\overleftarrow{W}})^{-1}\|\zeta\|^2 . \tag{5.11b}$$

Proof. The proof will be given only in the forward case and follows a short and well-known duality argument in Hilbert space theory. Fixing another element $\zeta' \in \mathbb{R}^N$, we have

$$\begin{aligned}
(Q_{\overrightarrow{W}}(\theta_{\zeta}) \,|\, \zeta') &= (\theta_{\zeta} \,|\, (Q_{\overrightarrow{W}})^{*}\zeta') \\
&= (((Q_{\overrightarrow{W}})^{*} \circ \sigma^{-1})(\zeta) \,|\, (Q_{\overrightarrow{W}})^{*}\zeta') \\
&= (\sigma^{-1}(\zeta) \,|\, Q_{\overrightarrow{W}} \circ (Q_{\overrightarrow{W}})^{*}\zeta') \\
&= (\sigma^{-1}\zeta \,|\, \sigma(\zeta')) \\
&= ((\sigma^{*} \circ \sigma^{-1})\zeta \,|\, \zeta') = (\zeta \,|\, \zeta'),
\end{aligned}$$

where the last equality is a consequence of the fact that $\sigma^{*} = \sigma$; therefore

$$(Q_{\overrightarrow{W}}(\theta_{\zeta}) - \zeta \,|\, \zeta') = 0, \quad \forall \zeta' ;$$

taking $\zeta' = \mathcal{Q}_{\overrightarrow{W}}(\theta_\zeta) - \zeta$ we get

$$\mathcal{Q}_{\overrightarrow{W}}(\theta_\zeta) - \zeta = 0 \,.$$

We proceed now to the proof of (5.10b):

$$(\theta_\zeta \,|\, \theta_\zeta)_{L^2} = \left((\mathcal{Q}_{\overrightarrow{W}})^* \circ \sigma^{-1}(\zeta) \,\middle|\, (\mathcal{Q}_{\overrightarrow{W}})^* \circ \sigma^{-1}(\zeta)\right)$$
$$= \left(\sigma^{-1}(\zeta) \,\middle|\, \mathcal{Q}_{\overrightarrow{W}} \circ (\mathcal{Q}_{\overrightarrow{W}})^* \circ \sigma^{-1}(\zeta)\right) = (\sigma^{-1}\zeta \,|\, \zeta). \quad \square$$

Proposition 5.3. *Assume that the vector fields A_k have bounded derivatives up to fourth order, then*

$$\sigma^{\rightarrow}, \ \sigma^{\leftarrow} \in D_1^p(\mathscr{W}^n) \,. \tag{5.12}$$

Proof. We have to show that $\mathcal{Q}^{\rightarrow} \in D_1^p(\mathscr{W}^n)$.

1) Fixing θ, we compute the derivatives

$$D_{\tau,k}(\mathcal{Q}_{\overrightarrow{W}})(\theta)$$

$$= \int_0^s dt \sum_{k=1}^n \left[\left(D_{\tau,k} J_{s \leftarrow t}^W\right)(A_k(r_W(t)))\,\theta^k(t) + J_{s \leftarrow t}^W\left(D_{\tau,k} A_k(r_W(t))\right)\theta^k(t)\right] \,.$$

Calculating first the second term, we get

$$D_{\tau,k} A_k(r_W(t)) = 1_{\{\tau < t\}}\, \mathbf{A}_k\, J_{t \leftarrow \tau}^W(A_k)$$

which is dominated by the bound on the derivatives of A_k. The computation of the first term involves the differentiation of the matrix-valued SDE (5.2):

$$d_s(J_{s \leftarrow t}^W) = \left(\sum_{k=1}^n \mathbf{A}_k(\zeta_W(t))\,dW_K(t) + \mathbf{A}_0(\zeta_W(t))\,dt\right) J_{s \leftarrow t}^W \,.$$

This derivative is obtained by differentiating again the coefficients $\mathbf{A}_k$ which leads to second derivatives of the vector fields A_k.

2) By means of (5.8) the derivative $D_{\tau,k}\sigma^{\xi\eta}$ is equal to

$$\int_0^s dt \sum_{k=1}^n (C_{\xi,\eta} + C_{\eta,\xi}), \quad \text{where}$$

$$C_{\xi,\eta} := \left(e^\xi \,\middle|\, D_{\tau,k}(J_{s \leftarrow t}^W A_k(r_W(t)))\right)_{\mathbb{R}^N} \left(e^\eta \,\middle|\, J_{s \leftarrow t}^W A_k(r_W(t))\right)_{\mathbb{R}^N} \,.$$

Note that the term $D_{\tau,k}(J_{s \leftarrow t}^W A_k(r_W(t)))$ has already been computed in the first step. $\square$

Remark 5.4. The computation of the derivatives of $\sigma^{\rightarrow}$ has been realized by solving linearized SDEs along the process; in principle, this procedure is implementable by a Monte-Carlo simulation.

5.3 Malliavin Covariance Matrix and Hörmander Bracket Conditions

Given two vector fields A_1, A_2 their *Lie bracket* is by definition the vector field C with the components:

$$C^\xi := \sum_{\eta \in \{1,\dots,N\}} \left(A_1^\eta \frac{\partial A_2}{\partial \xi^\eta} - A_2^\eta \frac{\partial A_1}{\partial \xi^\eta} \right), \quad \xi \in \{1,\dots,N\} . \tag{5.13}$$

The Lie bracket C of the vector fields A_1 and A_2 is denoted by $[A_1, A_2]$.

The *Lie algebra* $\mathcal{A}$ generated by n vector fields $A_1, \dots, A_n$ is defined as the vector space of all fields obtained as linear combinations with constant coefficients of the

$$A_k, \quad [A_k, A_\ell], \quad [[A_k, A_\ell], A_s], \quad [[[A_k, A_\ell], A_s], A_u], \quad \text{etc.}$$

Given $r \in \mathbb{R}^N$, let $\mathcal{A}(r) := \{ \zeta \in \mathbb{R}^N \mid \zeta = Z(r) \text{ for some } Z \in \mathcal{A} \}$.

Definition 5.5. *We say that vector fields $A_1, \dots, A_n$ satisfy the Hörmander criterion for hypoellipticity if $A_1, \dots, A_n$ are infinitely often differentiable and if*

$$\mathcal{A}(r) = \mathbb{R}^N \quad \text{for any } r \in \mathbb{R}^N . \tag{5.14}$$

Lemma 5.6 (Key lemma). *Assume that the vector fields A_k, along with their derivatives, are uniformly bounded and satisfy the Hörmander criterion for hypoellipticity. Denoting by $\lambda(W)$ the smallest eigenvalue of the covariance matrix (5.8), then*

$$\mathbb{E}\left[\lambda(W)^{-p} \right] < \infty, \quad \forall p < \infty . \tag{5.15}$$

Proof. See Malliavin [140] and Kusuoka–Stroock [124, 125]. □

Some aspects of the computation will be considered in the last section of this chapter.

5.4 Regularity by Predictable Smearing

We denote by $\pi_s(r_0, dr)$ the probability transition starting from r_0 at time 0 to be at time $s = T - t_0$ at the volume element dr.

Given $p > 1$, we say that $(p, 1)$-*forward regularity* holds true if the following formula of integration by parts is satisfied: $\forall \phi \in C_b^1$ and $\forall \xi \in \{1, \dots, N\}$,

$$\int_{\mathbb{R}^N} \frac{\partial \phi}{\partial r^\xi}(r) \, \pi_s(r_0, dr) = \int_{\mathbb{R}^N} \phi(r) \, K_\xi(r) \, \pi_s(r_0, dr) , \tag{5.16}$$

where $\int |K_\xi(r)|^p \, \pi_s(r_0, dr) < \infty$.

It is obvious that conditions are needed on the driving vector fields A_k in order that forward regularity can hold true; for instance, let A_1, A_2 be the two first coordinate vector fields of $\mathbb{R}^N$, take $A_0 = 0$, then the generated diffusion is a Brownian motion on $\mathbb{R}^2$ and it is impossible to find an integration by parts formula for other coordinate vector fields.

Given $p > 1$, we say that $(p, 1)$-*backward regularity* holds if for any $\phi \in C_b^1$ and any $\xi \in \{1, \ldots, N\}$,

$$\lim_{\varepsilon \to 0} \frac{1}{\varepsilon} \int_{\mathbb{R}^N} \left(\pi_s(r_0 + \varepsilon e^\xi, dr) - \pi_s(r_0, dr) \right) \phi(r) = \int_{\mathbb{R}^N} \phi(r) \, H_\xi(r) \, \pi_s(r_0, dr) ,$$

$$(5.17)$$

where $\int |H_\xi(r)|^p \, \pi_s(r_0, dr) < \infty$.

Theorem 5.7. *Let the Hörmander hypoellipticity criterion (5.14) be satisfied, and assume that derivatives of any order of the vector fields A_k are uniformly bounded. Then backward and forward regularity hold true for every $p > 1$ and every $r_0 \in \mathbb{R}^N$.*

Proof. First we are going to prove backward regularity. Consider the coordinate vector fields $(e^\eta)_{1 \leq \eta \leq N}$, and for $\eta \in \{1, \ldots, N\}$ define vector fields Y_η on $\mathscr{W}^n$ by

$$Y_\eta(W) := (\mathcal{Q}_{\overleftarrow{W}})^*(e'') .$$

$$(5.18)$$

It results from the backward structure that each Y_η is *predictable*; therefore the divergence of Y_η is computable by the following Itô integral:

$$\vartheta(Y_\eta) = \sum_{k=1}^n \int_0^s Y_\eta^k(\tau) \, dW_k(\tau) .$$

$$(5.19)$$

Denote $\gamma_{\overleftarrow{W}} := (\sigma_{\overleftarrow{W}})^{-1}$ and consider the vector field Z_ξ on $\mathscr{W}^n$ defined by the following linear combination of the vector fields Y_η:

$$Z_\xi := \sum_\eta (\gamma_{\overleftarrow{W}})_\xi^\eta Y_\eta; \quad \text{then } \vartheta(Z_\xi) = \sum_\eta \left((\gamma_{\overleftarrow{w}})_\xi^\eta \vartheta(Y_\eta) - D_{Y_\eta}(\gamma_{\overleftarrow{w}})_\xi^\eta \right) . \quad (5.20)$$

According to (5.16) and (1.19), the divergence $\vartheta(Z)$ exists and $\vartheta(Z) \in L^p$. Using now (5.11a) gives

$$\mathbb{E}\left[\vartheta(Z_\xi) \, \phi(r_W(s))\right] = \mathbb{E}\left[D_{Z_\xi}(\phi(r_W(s)))\right]$$

$$= \lim_{\varepsilon \to 0} \frac{1}{\varepsilon} \int_{\mathbb{R}^N} \left(\pi_s(r_0 + \varepsilon e^\xi, dr) - \pi_s(r_0, dr) \right) \phi(r). \quad (5.21)$$

Finally denote by $\mathcal{E}_s$ the σ-field generated by $W \mapsto r_W(s)$. Taking conditional expectation of (5.21) we get

$$\int_{\mathbb{R}^N} \mathbb{E}^{\mathcal{E}_s}[\vartheta(Z_\xi)] \, \phi(r) \, \pi_s(r_0, dr)$$

$$= \lim_{\varepsilon \to 0} \frac{1}{\varepsilon} \int_{\mathbb{R}^N} \left(\pi_s(r_0 + \varepsilon e^\xi, dr) - \pi_s(r_0, dr) \right) \phi(r) .$$

To prove forward regularity we consider the conjugated matrix $J_{0 \leftarrow s}^W =: \delta_W$. To a given coordinate vector field e^ξ on $\mathbb{R}^N$, we define a vector field U_ξ on $\mathscr{W}^n$ by the following linear combination of the vector fields Z_*:

$$U_\xi = \sum_\eta (\delta_W)_\xi^\eta Z_\eta; \quad \text{then } \vartheta(U_\xi) = \sum_\eta \left((\delta_W)_\xi^\eta \vartheta(Z_\eta) - D_{Z_\eta}(\delta_W)_\xi^\eta \right). \quad (5.22)$$

Then

$$\mathbb{E}\left[\vartheta(U_\xi) \, \phi(r_W(s)) \right] = \mathbb{E}\left[D_{U_\xi}(\phi(r_W(s))) \right] = \int_{\mathbb{R}^N} \frac{\partial \phi}{\partial r^\xi}(r) \, \pi_s(r_0, dr). \quad \square$$

Remark 5.8. The experienced reader will see that the same result holds true if invertibility of $\sigma_{\overleftarrow{W}}$ is replaced by the condition that ζ lies in the range of $\sigma_{\overleftarrow{W}}$, where ζ denotes a tangent vector at the initial point r_0.

Remark 5.9. The approach used in this section is implementable in a Monte-Carlo simulation.

5.5 Forward Regularity by an Infinite-Dimensional Heat Equation

The idea is to use the Brownian sheets to construct heat processes on $\mathscr{W}^n$. We use the following representation of the Brownian sheet

$$W_s(\tau) = \tau B_0(s) + \sqrt{2} \sum_{q=1}^\infty \frac{\sin(q\pi\tau)}{q} B_q(s), \quad \tau \in [0,1], \quad (5.23)$$

where $\{B_q\}_{q \geq 0}$ is an infinite sequence of independent $\mathbb{R}^n$-valued Brownian motions satisfying $B_q(0) = 0$.

We fix s and consider the SDE, companion of (5.1), taking now the form

$$d_\tau(r_{W_s}(\tau)) = \sum_{k=1}^n A_k(r_{W_s}(\tau)) \, dW_s^k(\tau) + A_0(r_{W_s}(\tau)) \, d\tau, \quad r_{W_s}(0) = r_0.$$

$$(5.24)$$

We consider the functional $\Phi : \{B_\bullet\} \to \mathbb{R}^N$ defined as $\Phi(B_\bullet(s)) = r_{W_s}(1)$.

Theorem 5.10. *The process $s \mapsto \Phi(B_\bullet(s))$ is an $\mathbb{R}^N$-valued semimartingale such that*

$$(\sigma_{\overrightarrow{W_s}})^{\xi,\eta} = \lim_{\varepsilon \to 0} \frac{1}{\varepsilon} \mathbb{E}^{\mathcal{N}_s} \Big[\left(e^\xi \mid \Phi(B_\bullet(s+\varepsilon)) - \Phi(B_\bullet(s)) \right)$$

$$\times \left(e^\eta \mid \Phi(B_\bullet(s+\varepsilon)) - \Phi(B_\bullet(s)) \right) \Big]. \quad (5.25)$$

There exists a vector field $\alpha(r,s) = (\alpha^\xi(r,s) : 1 \leq \xi \leq N)$ such that in terms of the time-dependent elliptic operator

$$\mathscr{L}_s = \frac{1}{2} \sum_{\xi,\eta} \bar{\sigma}^{\xi,\eta}(r,s) \frac{\partial^2}{\partial r^\xi \partial r^\eta} + \sum_\xi \alpha^\xi(r,s) \frac{\partial}{\partial r^\xi}, \quad \bar{\sigma}(r,s) := \mathbb{E}^{\Phi(B.(s))=r}[\sigma_{\overrightarrow{W}_s}],$$

the following formula holds:

$$\frac{\partial}{\partial s} \int_{\mathbb{R}^N} \pi_s(r_0, dr) \, \phi(r) = \int_{\mathbb{R}^N} \pi_s(r_0, dr) \, (\mathscr{L}_s \phi)(r), \quad \forall \phi. \tag{5.26}$$

Proof. Use Itô calculus for proving (5.25). □

Remark 5.11. This methodology reduces the study of a hypoelliptic operator to the study of a non-autonomous elliptic operator $\mathscr{L}_s$. It is of interest to see how the elliptic symbol varies. Itô calculus applied to the process $s \mapsto \sigma_{\overrightarrow{W}_s}$ will give rise to a matrix-valued SDE for $d_s(\sigma_{\overrightarrow{W}_s})$.

5.6 Instability of Hedging Digital Options in HJM Models

We follow the Musiela parametrization of the Heath–Jarrow–Morton model (otherwise HJM model). In this parametrization a market state at time t_0 is described by the instantaneous interest rate $r_{t_0}(\xi)$ to a maturity ξ, defined for $\xi \in \mathbb{R}^+$. The price $P(t_0, T)$ of a default-free bond traded at t_0 for a maturity at $t_0 + T$ then has the value

$$P(t_0, T) = \exp\left(\int_0^T r_{t_0}(\xi) \, d\xi \right).$$

Now the following question can be asked: what is the most general model giving rise to a continuous arbitrage-free evolution? The HJM model answers this general question. We shall limit the generality of the HJM model to the case of a finite number of scalar-valued driving Brownian motions $W_1, \ldots, W_n$. Then the HJM model is expressed by the following SDE: $\forall \xi \in \mathbb{R}^+$,

$$dr_t(\xi) = \sum_{k=1}^n B_k(t, \xi) \, dW_k(t) + \left\{ \frac{\partial r_t}{\partial \xi}(\xi) + \sum_{k=1}^n \int_0^\xi B_k(t, \xi) B_k(t, \eta) \, d\eta \right\} dt.$$

$$\tag{5.27}$$

As in the case of an elliptic market, the drift of the risk-free process is completely determined by the volatility matrix $B.(t, \cdot)$.

The hypothesis of *market completeness*, in the sense that exogenous factors of stochasticity are negligible, assumes that the stochasticity matrix $A.(t, \cdot)$ is $\mathcal{N}_t$ measurable where $\mathcal{N}_t$ denotes the σ-field generated by $r_s(\cdot)$, $s \leq t$. We shall work under the hypothesis of Markovian completeness which means that this dependence is factorizable through the final value $r_t(\cdot)$. More precisely, *Markovian completeness* means that there exists n such that

$$dr_t(\xi) = \sum_{k=1}^{n} (A_k(r_t))(\xi) \, dW_k(t)$$

$$+ \left\{ \frac{\partial r_t}{\partial \xi}(\xi) + \sum_{k=1}^{n} \int_0^{\xi} (A_k(r_t))(\xi) \, (A_k(r_t))(\eta) \, d\eta \right\} dt. \qquad (5.28)$$

An appropriate notion of "smoothness" of vector fields is a necessary hypothesis in order to prove existence and uniqueness of solutions. Banach space type differential calculus is not suitable because the operator $r \mapsto \partial r_t / \partial \xi$ is not bounded. To this end, a theory of differential calculus in Fréchet space is needed (see Filipović–Teichmann [72, 73]). A precise differential calculus treatment of (5.28) is in itself a whole theory which we shall only touch in this book.

Our point is to avoid functional analysis and to concentrate on proving that for reasonable finite-dimensional approximations *numerical instabilities appear in digital option hedging.*

Let $C = C([0, \infty[)$ be the space of continuous functions on $[0, \infty[$. Define the *logarithmic derivative* of a measure μ_λ depending on a parameter λ as the function satisfying, for every test function f,

$$\partial_\lambda \int_C f(r) \, \mu_\lambda(dr) = \int_C f(r) \, (\partial_\lambda \log(\mu_\lambda))(r) \, \mu_\lambda(dr) \, .$$

Then a Hilbert norm on tangent vectors at r_0 is given by

$$\|z\|_t^2 = \int_C \left[\frac{d}{d\varepsilon} \Big|_{\varepsilon=0} \log(\pi_t(r_0 + \varepsilon z, \cdot))(r) \right]^2 \pi_t(r_0, dr) \, . \qquad (5.29)$$

Definition of Compartmentation (Hypoelliptic Global Version)

$$\text{The two norms } \|z\|_s, \|z\|_{s'}, \ s \neq s' \text{ are inequivalent.} \qquad (5.30)$$

The operator $\mathcal{Q}_W^{\leftarrow}$ has been defined in (5.3). We choose now a Hilbert metric $\|\cdot\|_C$ on C. Using the canonical Hilbert space structure of $L^2([0, s]; \mathbb{R}^n)$, the adjoint $(\mathcal{Q}_{W_s}^{\leftarrow})^*$ gives the operator

$$\sigma_{W_s}^{\leftarrow} = Q_{W_s}^{\leftarrow} \circ (Q_{W_s}^{\leftarrow})^* \, .$$

Theorem 5.12. *We have the estimate*

$$\|y\|_s^2 \leq \inf \left\{ \|u\|_C : \ \sigma_{W_s}^{\leftarrow}(u) = y \right\} \, .$$

Proof. See Theorem 5.7. □

Theorem 5.13 (Pathwise compartmentation principle). *Let W be a fixed trajectory of the Brownian motion and denote by $\sigma_W^{\leftarrow}(s)$ the covariance matrix computed on $[0, s]$. Then, generically, the range of $\sigma_W^{\leftarrow}(s)$ is strictly increasing as a function of s.*

Proof. See [5]. □

Corollary 5.14. *There exist digital options for which the Ocone–Karatzas hedging formula becomes unstable.*

See Baudoin–Teichmann [24] for related results.

5.7 Econometric Observation of an Interest Rate Market

From our point of view, an *econometric computation* will be a numerical computation realized in real time and made from observations of a single time series of market data, without any quantitative assumption or any choice of model.

Firstly we can compute historical variance from the market evolution; for this purpose we can use the methodology of Fourier series described in Appendix A or other algorithms (e.g., see [92]).

We get a time-dependent $N \times N$ covariance matrix $d_t r(t) * d_t r(t) =: \mathcal{C}(t) \, dt$; denote by $\lambda_{k,t}$ its eigenvalues and by $\tilde{\phi}_{k,t}$ the corresponding normalized eigenfunctions. We assume that only a relatively small number N_0 of eigenvalues are not very close to zero. Denoting by $\phi_{k,t} = \sqrt{\lambda_{k,t}} \, \tilde{\phi}_{k,t}$ the *canonical eigenvectors*, we have

$$\mathcal{C}(t) = \sum_{k=1}^{N_0} \phi_{k,t} \otimes \phi_{k,t} .$$

The vector-valued function $t \mapsto \phi_{k,t}$ is the econometric reconstruction of the vector-valued function $t \mapsto A_k(r_W(t))$ under the assumption that model (5.1) holds true.

We keep the hypotheses used in Chap. 3 for constructing the feedback volatility rate. We assume that

$$\phi_{k,t} = \Phi_k(r_W(t)) \tag{5.31}$$

where the Φ_k are smooth unknown functions of r varying slowly in time. In a real market the Φ_k depend also on external factors; as our purpose is to study the market on a short period of time, assumption (5.31) appears reasonable.

We use a model-free approach and make no assumption on the actual expression of the functions Φ_k. The main fact is that, using the methodology of *iterated volatilities*, we are able to compute econometrically pathwise brackets $[\Phi_k, \Phi_l]$.

Theorem 5.15. *Under assumption* (5.31) *we have*

$$\partial_{\phi_{s,t}}(\phi_{q,t}^\alpha) \, dt = \frac{1}{\lambda_{s,t}} \sum_{\beta=1}^{N} \phi_{s,t}^\beta \left(d_t \phi_{q,t}^\alpha * d_t r_t^\beta \right) . \tag{5.32}$$

Proof. From our point of view, it is legitimate to compute cross-covariances of any processes defined on the econometric time evolution. Under assumption (5.31), we have

$$d_t \phi_{q,t}^\alpha * dr_t^\beta = \sum_{\gamma=1}^N \mathcal{C}_{\gamma,\beta}\, \partial_\gamma \Phi_q^\alpha\, dt \,.$$

We multiply this identity by $\phi_{s,t}^\beta$ and sum on β:

$$\sum_\beta \phi_{s,t}^\beta \left(d_t \phi_{q,t}^\alpha * d_t r_t^\beta \right) = \sum_\gamma \left(\sum_\beta \phi_{s,t}^\beta \mathcal{C}_{\gamma,\beta} \right) \partial_\gamma \phi_{q,t}^\alpha\, dt = \lambda_{s,t} \partial_{\phi_{s,t}} (\phi_{q,t}^\alpha)\, dt \,.$$

Corollary 5.16. *The brackets of the vector fields Φ_k are given by the following expressions:*

$$[\Phi_s, \Phi_k]^\alpha\, dt = \frac{1}{\lambda_{s,t}} \sum_\beta \phi_{s,t}^\beta \left(d_t \phi_{k,t}^\alpha * d_t r_t^\beta \right) - \frac{1}{\lambda_{k,t}} \sum_\beta \phi_{k,t}^\beta \left(d_t \phi_{s,t}^\alpha * d_t r_t^\beta \right) \,. \quad (5.33)$$

All previous statements are mathematical statements for a market driven by the SDE (5.1). It is possible to use these statements as econometric tools to decipher the market state if, at an appropriated time scale, the vectors $\phi_{k,t}$, $[\phi_{k,t}, \phi_{\ell,t}]$ show some kind of stability.

6

Insider Trading

Anticipative stochastic calculus can be considered as a major progress of Stochastic Analysis in the decade 1980–1990: starting with [82] and [160] and developed in [46, 75, 97, 98, 103, 157, 159, 206].

Application of anticipative calculus to insider behaviour began with [89, 90], and has been developed to an elaborated theory in [6, 54, 99, 100, 131]. This chapter is mainly devoted to a presentation of work of P. Imkeller and his co-workers [6, 54, 75, 97–100].

6.1 A Toy Model: the Brownian Bridge

As this chapter deals with abstract concepts, we shall first present in this section the relevant concepts in the concrete framework of the *Brownian bridge*.

Fixing two points $a, b \in \mathbb{R}^n$, let $C_T^{a \to b}$ denote the affine space of continuous functions $u \colon [0, T] \to \mathbb{R}^n$ such that $u(0) = a$ and $u(T) = b$. Given a positive integer s, let $\xi_k = k 2^{-s} T$ for $k = 0, 1, \ldots, 2^s$ be the dyadic partition of the interval $[0, T]$. As in Sect. 1.2, we consider the subspace ${}^s C_T^{a \to b}$ of $C_T^{a \to b}$ consisting of the maps which are linear on the sub-intervals between consecutive points of the dyadic partition.

We define a probability measure ${}^s \rho_T^{a \to b}$ on ${}^s C_T^{a \to b}$ by the formula

$$\frac{\pi_\varepsilon(a, \eta_1)\, \pi_\varepsilon(\eta_1, \eta_2) \ldots \pi_\varepsilon(\eta_{2^{-s}-2}, \eta_{2^{-s}-1})\, \pi_\varepsilon(\eta_{2^{-s}-1}, b)}{\pi_T(a, b)} \bigotimes_{k=1}^{2^s - 1} d\eta_k$$

where $\varepsilon = T 2^{-s}$ and where π_t is the heat kernel associated to Brownian motion on $\mathbb{R}^n$:

$$\pi_t(\xi, \xi') = \frac{1}{(2\pi t)^{n/2}} \exp\left(-\frac{\|\xi - \xi'\|^2}{2t}\right).$$

Theorem 6.1. *As $s \to \infty$, the measures ${}^s\rho_T^{a \to b}$ converge weakly to a Borel measure $\rho_T^{a \to b}$ carried by $C_T^{a \to b}$. Given $0 < t_1 < \ldots < t_q < 1$, the image of $\rho_T^{a \to b}$ under the evaluations $\{e_{t_i}\}_{i=1,\ldots,q}$ is*

$$\frac{\pi_{t_1}(a, \eta_1)\,\pi_{t_2-t_1}(\eta_1, \eta_2) \ldots \pi_{t_q-t_{q-1}}(\eta_{q-1}, \eta_q)\,\pi_{1-t_q}(\eta_q, b)}{\pi_T(a, b)} \bigotimes_{k=1}^{q} d\eta_k \,. \qquad (6.1)$$

The measure $\rho_1^{0 \to b}$ is the conditional law of the Brownian motion W indexed by $[0, 1]$, under the conditioning $W(1) = b$, or equivalently:

$$\mathbb{E}[\Phi(W)] = \int_{\mathscr{W}^n} \Phi \, d\gamma = \int_{\mathbb{R}^n} \pi_1(0, b) \, db \left[\int_{C_1^{0 \to b}} \Phi \, d\rho_1^{0 \to b} \right]. \qquad (6.2)$$

Proof. The first statement is proved following the lines of the proof to Theorem 1.6; the proof shows at the same time that $\rho_T^{a \to b}$ is supported by the space of Hölder continuous paths of exponent α where $\alpha < 1/2$. Formula (6.1) is first checked in the case where the t_k are dyadic fractions; the general case is obtained by passing to the limit. Note that formula (6.2) is equivalent to the analogous statement for the image measures under the evaluation map $\{e_{t_i}\}_{i=1,\ldots,q}$. The image of the Wiener measure γ by the evaluation map is

$$\left(\pi_{t_1}(a, \eta_1)\,\pi_{t_2-t_1}(\eta_1, \eta_2) \ldots \pi_{t_q-t_{q-1}}(\eta_{q-1}, \eta_q)\,\pi_{1-t_q}(\eta_q, b) \bigotimes_{k=1}^{q} d\eta_k \right) db \,.$$

This completes the proof of the theorem. □

Theorem 6.2. [68] *Consider the* SDE

$$dB(t) = d\tilde{W}(t) + \mathcal{I}_t(B(t))\, dt, \quad \mathcal{I}_t(\xi) := \nabla \log \pi_{1-t}(\cdot, b)(\xi) \,, \qquad (6.3)$$

where $\tilde{W}$ is a Brownian motion on $\mathbb{R}^n$. The measure $\rho_1^{0 \to b}$ is the law of the process $B(t)$, $0 \le t < 1$, with initial condition $B(0) = a$.

The time-dependent vector field $\mathcal{I}_t$ defined in (6.3) is called the *information drift*.

Proof. Let

$$\Delta = \frac{1}{2} \sum_{k=1}^{n} \frac{\partial^2}{\partial \xi_k^2}$$

be the Laplacian on $\mathbb{R}^n$ and

$$\mathscr{L} = \Delta + \frac{\partial}{\partial t}$$

the corresponding heat operator. Consider the parabolic operator

$$\mathscr{S} = \mathscr{L} + \mathcal{I} * \nabla \,. \qquad (6.4)$$

As $\mathscr{L}v = 0$ for $v(\xi, t) := \pi_{1-t}(\xi, b)$, we get the Doob intertwining property

$$\mathscr{S}(\phi) = \frac{1}{v}\mathscr{L}(v\phi) . \tag{6.5}$$

This intertwining of infinitesimal generators extends by exponentiation to the intertwining of the associated semigroups on measures:

$$U^{\mathscr{S}}_{t_1 \to t_2} = \frac{1}{v} \circ U^{\mathscr{L}}_{t_1 \to t_2} \circ v, \quad 0 < t_1 < t_2 < 1 .$$

The last relation coincides with (6.1). □

Theorem 6.3. *For $0 \le t < 1$ let $\mathscr{F}_t$ be the filtration generated by $B(s)$, $s \le t$; denote by γ the Wiener measure. Then the Radon–Nikodym derivative is given by*

$$\left.\frac{d\rho_1^{0 \to b}}{d\gamma}\right|_{\mathscr{F}_t} = \exp\left(-\int_0^t \sum_{k=1}^n \mathcal{I}_s^k(B(s))\, d\tilde{W}_k(s) - \frac{1}{2}\int_0^t \left\| \mathcal{I}_s(B(s)) \right\|_{\mathbb{R}^n}^2 ds \right) . \tag{6.6}$$

Proof. The proof is performed by a Girsanov transformation on the SDE (6.3), see Theorem 1.23. □

6.2 Information Drift and Stochastic Calculus of Variations

In order to simplify the notation we confine ourselves to a market with a single risky asset whose price S is given by a semimartingale on the Wiener space $\mathscr{W}$. More formally, S is assumed to be driven by the stochastic differential equation

$$dS_W(t) = \beta_t\, dW(t) + \alpha_t\, dt \tag{6.7}$$

where W is a one-dimensional Brownian motion and β_t, α_t are $\mathscr{F}_t$-measurable; $(\mathscr{F}_t)$ denotes the Brownian filtration generated by W.

The regular trader is operating on the market at time t, only with the knowledge of $\mathscr{F}_t$ at hand. The trading operations are done on the interval $[0, T]$. We assume existence of a random variable G which is $\mathscr{F}_T$-measurable for the regular trader. For the insider however the value of G is known from the beginning of the trading period. We denote by $\mathscr{G}_t$ the *insider filtration* $\mathscr{F}_t \vee \sigma(G)$; any $\mathscr{G}_t$-measurable portfolio is an admissible strategy for the insider.

The *information drift* is defined as the $\mathscr{G}_t$-measurable function $\mathcal{I}_t$ such that

$$W'(t) := W(t) - \int_0^t \mathcal{I}_s \, ds \text{ is a } (\mathscr{G}_t)\text{-martingale.} \tag{6.8}$$

In this section we use Stochastic Analysis on the space $\mathscr{W}$ with respect to the filtration $(\mathscr{F}_t)$. The *G-conditional law process* $t \mapsto \mu_{t,W}$ takes its values in the simplex of probability measures on $\mathbb{R}$ and is defined by

$$\mathbb{E}^{\mathscr{F}_t}[\phi(G)] = \int_{\mathbb{R}} \phi(\xi) \, \mu_{t,W}(d\xi) \,. \tag{6.9}$$

At maturity $t = T$ the conditional law process takes its values in the Dirac measures; the value at time $t = 0$ is the law of G.

Proposition 6.4. *The G-conditional law process is an $(\mathscr{F}_t)$-martingale.*

Remark 6.5. Note that in a simplex it is legitimate to take barycentres of a collection of points; thus the above statement is meaningful.

Proof. Given $t < t' < T$, we compute

$$\langle \phi, \mathbb{E}^{\mathscr{F}_t}[\mu_{t',W}] \rangle = \mathbb{E}^{\mathscr{F}_t}[\langle \phi, \mu_{t',w} \rangle]$$
$$= \mathbb{E}^{\mathscr{F}_t}[\mathbb{E}^{\mathscr{F}_{t'}}(\phi(G))]$$
$$= \mathbb{E}^{\mathscr{F}_t}[\phi(G)] = \langle \phi, \mu_{t,W} \rangle. \quad \square$$

Every scalar-valued martingale on $\mathscr{W}$ can be written as a stochastic integral; by the next condition we assume that the same holds true for the conditional law process:

1. There exists an adapted process $a_W(t)$ taking values in the signed measures such that $\forall h \in C_b(\mathbb{R})$ the following representation holds:

$$\langle h, \mu_{t,W} - \mu_0 \rangle = \int_0^t \langle h, a_W(s) \rangle \, dW(s) \,. \tag{6.10}$$

2. For any $s < 1$, $a_W(s)$ is absolutely continuous with respect to $\mu_{s,W}$:

$$\frac{da_W(s)}{d\mu_{s,W}} =: \delta_{s,W} \,. \tag{6.11}$$

3. Finally, assume that

$$\forall t < T, \quad \mathbb{E}\left[\int_0^t |\delta_{s,W}| \, ds\right] < \infty \,. \tag{6.12}$$

Theorem 6.6. *Suppose that the conditions $1, 2$ and 3 above hold true. Then the information drift $\mathcal{I}$ exists and equals*

$$\mathcal{I}_t = \delta_{t,W}(\xi_0) \tag{6.13}$$

where ξ_0 is the value of G known by the insider.

Proof. Consider a test function h and compute for $\tau > t$

$$A := \mathbb{E}^{\mathscr{F}_t}\Big[h(G)(W(\tau) - W(t))\Big]$$
$$= \mathbb{E}^{\mathscr{F}_t}\Big[(W(\tau) - W(t))\int_{\mathbb{R}} h(\xi)\,\mu_{\tau,W}(d\xi)\Big]. \qquad (6.14)$$

Writing

$$\langle h, \mu_{\tau,W}\rangle = \langle h, \mu_{t,W}\rangle + \int_t^\tau \langle h, a_{s,W}\rangle\,dW(s)\,,$$

the right-hand-side of (6.14) gives a sum of two terms. The first term is easily seen to vanish:

$$\mathbb{E}^{\mathscr{F}_t}\Big[(W(\tau) - W(t))\int_{\mathbb{R}} h(\xi)\,\mu_{t,W}(d\xi)\Big]$$
$$= \Big(\int h(\xi)\,\mu_{t,W}(d\xi)\Big)\,\mathbb{E}^{\mathscr{F}_t}[W(\tau) - W(t)] = 0\,.$$

For the second term we note that

$$\mathbb{E}^{\mathscr{F}_t}\Big[\int_t^\tau\Big(\int_{\mathbb{R}} h(\xi)\,a_{s,W}(d\xi)\Big)\,ds\Big] = \mathbb{E}^{\mathscr{F}_t}\Big[\int_t^\tau\Big(\int h(\zeta)\,\delta_{s,W}(\zeta)\,\mu_{s,W}(d\xi)\Big)\,ds\Big]\,.$$

Letting $\Psi_s(\xi) := \delta_{s,W}(\xi)h(\xi)$, we have

$$\int h(\xi)\delta_{s,W}(\xi)\,\mu_{s,W}(d\xi) = \int \Psi_s(\xi)\,\mu_{s,W}(d\xi)$$
$$= \mathbb{E}^{\mathscr{F}_s}[\Psi(G)] = \mathbb{E}^{\mathscr{F}_s}[h(G)\,\delta_{s,W}(G)]\,.$$

Therefore

$$A = \mathbb{E}^{\mathscr{F}_t}\Big[h(G)\int_t^\tau \delta_{s,W}(G)\,ds\Big]\,,$$

and we deduce that

$$\mathbb{E}^{\mathscr{F}_t}\Big[h(G)\Big(W(\tau) - W(t) - \int_t^\tau \delta_{s,W}(G)\,ds\Big)\Big] = 0. \qquad \Box$$

6.3 Integral Representation of Measure-Valued Martingales

The purpose of this section is to prove that condition (6.10) holds true under a very weak hypothesis of differentiability. The reader interested mainly in properties of the information drift can skip this section and proceed to Sect. 6.4.

The main technical tool in this section is the stochastic calculus of variation for measure-valued functionals. Denote by $\mathscr{M}$ the vector space of signed Borel measures on $\mathbb{R}$ of finite total variation; hence if $|\mu|$ denotes the total variation norm of $\mu \in \mathcal{M}$, then $|\mu| < \infty$. Let $C_b(\mathbb{R})$ be the space of bounded continuous functions on $\mathbb{R}$ which is dual to $\mathcal{M}$ through the pairing $\langle f, \mu \rangle = \int f \, d\mu$.

Note that we can find a sequence $\{f_i\}$ in $C_b(\mathbb{R})$ such that $\|f_i\|_{C_b(\mathbb{R})} = 1$ and such that $\sup_i |\langle f_i, \mu \rangle| = |\mu|$ for all $\mu \in \mathcal{M}$. We define a mapping

$$\Phi \colon \mathcal{M} \to \mathbb{R}^{\mathbb{N}}, \quad \Phi^i(\mu) := \langle f_i, \mu \rangle .$$

We consider $\mathcal{M}$-valued functionals $\psi : W \mapsto \mu_W \in \mathcal{M}$ and define

$$D_1^p(\mathcal{W}; \mathcal{M}) = \left\{ \psi : \ \Phi^i \circ \psi \in D_1^p(\mathcal{W}), \ i = 1, \ldots, N, \ \text{and} \ \|\psi\|_{D_1^p}^p < \infty \right\},$$

where

$$\|\psi\|_{D_1^p}^p := \mathbb{E}\left[|\psi|^p + \left(\int_0^1 |D_t \psi|^2 \, dt \right)^{p/2} \right] \tag{6.15}$$

and where $|D_t \psi| := \sup_i |D_t(\Phi^i \circ \psi)|$. Then a.s. in W, there exists $c(W, \cdot) \in L^2([0,1])$ such that

$$|D_t(\Phi^i \circ \psi)| \le c(W, t) .$$

This domination implies that:

$$\exists \nu_{t,W} \in \mathcal{M} \ \text{such that} \ D_t(\Phi^i \circ \psi) = \langle f_i, \nu_{t,W} \rangle; \quad (D_t \psi)(W) := \nu_{t,W} . \tag{6.16}$$

Theorem 6.7 (Main theorem). *Assume that the conditional law process satisfies*

$$\mu_{1-\varepsilon, \cdot} \in D_1^p(\mathcal{W}; \mathcal{M}), \quad \forall \varepsilon > 0 . \tag{6.17}$$

Then (6.10) holds true. If furthermore (6.11) and (6.12) are satisfied, the information drift $\mathcal{I}_s$ is well-defined for $s < 1$.

Proof. Fix $\varepsilon > 0$ and let $\psi^i(W) = \Phi^i(\mu_{1-\varepsilon, W})$; then the Ocone–Karatzas formula can be applied on $[0, 1 - \varepsilon]$:

$$\psi^i - \mu^i = \int_0^{1-\varepsilon} \mathbb{E}^{\mathscr{F}_s}[D_s \psi^i] \, dW(s), \quad \mu^i := \langle f^i, \mu \rangle .$$

The same formula holds for any finite linear combination of the f_i. Fixing $h \in C_b(\mathbb{R})$, we find a sequence g_k of linear combinations of the f_i such that

$$\lim g_k(\xi) = h(\xi), \ \text{uniformly} \ \xi \in [-A, A], \quad \forall A < \infty , \tag{6.18}$$

$$\text{and} \quad \sup_k \|g_k\|_{C_b(\mathbb{R})} < \infty .$$

Then

$$\langle g_k, \mu_{1-\varepsilon} \rangle - \langle g_k, \mu \rangle = \int_0^{1-\varepsilon} \langle g_k, \mathbb{E}^{\mathscr{F}_t}[D_t \psi] \rangle \, dW(t)$$

where $\psi := \mu_{1-\varepsilon, W}$ and where $D_t \psi$ is defined in (6.16); letting $k \to \infty$ and using (6.18) we get (6.10). $\square$

6.4 Insider Additional Utility

Assume that the price of the risky asset is given by the SDE

$$dS_t = S_t(\sigma_t \, dW(t) + \alpha_t \, dt)$$

where the coefficients α_t, σ_t are $\mathscr{F}_t$-measurable and where $\sigma_t \geq \sigma > 0$.

The regular trader portfolio is given by an $\mathscr{F}_t$-measurable function π_t describing the quantity of the risky asset. Then the value $V(t)$ of the portfolio at time t is given by the SDE

$$dV(t) = V(t)\pi_t \, dS(t) \ .$$

This linear SDE can be explicitly integrated and we get

$$\frac{V(t)}{V(0)} = \exp\left(\int_0^t \sigma_s \, dW(s) - \int_0^t \left(\frac{1}{2}\pi^2\sigma_s^2 - \pi_s\alpha_s \right) ds \right) \ .$$

Choosing as *utility function* $U(t) = \log V(t)$, we get

$$\mathbb{E}[U(t)] = \mathbb{E}\left[\int_0^t \left(\pi_s\alpha_s - \frac{1}{2}\pi_s^2\sigma_s^2 \right) ds \right] + \log V(0) \ . \tag{6.19}$$

Proposition 6.8. *The portfolio maximizing the expectation of the logarithmic utility of the regular trader is given by $\pi_s = \alpha_s/\sigma_s^2$; the maximum expected utility equals*

$$U^r = \frac{1}{2} \mathbb{E}\left[\int_0^t \frac{\alpha_s^2}{\sigma_s^2} ds \right] \ . \tag{6.20}$$

Remark 6.9. Negative values of π_s are admissible; negative portfolios are realized on the market by selling call options.

Proof. As a consequence of the hypotheses made on σ_s and α_s, the proposed extremal portfolio is $(\mathscr{F}_s)$-measurable. For fixed s it realizes the maximum of the quadratic form $-\frac{1}{2}\pi^2\sigma_s^2 + \pi\alpha_s$. □

Insider Portfolio

With respect to the insider filtration $(\mathscr{G}_t)$ the Brownian motion W becomes a semimartingale with drift; the drift is given by the information drift:

$$dW = dW' + \mathcal{I}dt \ , \tag{6.21}$$

where W' is a $(\mathscr{G}_t)$-Brownian motion. Note that $\mathcal{I}_t$ is $\mathscr{G}_t$-measurable.

Theorem 6.10. *The portfolio maximizing the expected utility of the insider is*

$$\pi_s = \frac{\alpha_s + \mathcal{I}_s}{\sigma_s^2}$$

with $U^i = U^r + U^a$ as the maximal expected utility and the additional utility U^a given by

$$U^a := \frac{1}{2} \mathbb{E}\left[\int_0^t \frac{\mathcal{I}_s^2}{\sigma_s^2}\, ds\right]. \tag{6.22}$$

Proof. Using (6.21), we resume the computations made for the regular trader. The situation is reduced to the previous situation with $\alpha_s \mapsto \alpha_s + \mathcal{I}_s$, and we get

$$U^i = \frac{1}{2} \mathbb{E}\left[\int_0^t \frac{(\alpha_s + \mathcal{I}_s)^2}{\sigma_s^2}\, ds\right].$$

Formula (6.22) is equivalent to the following orthogonality relation:

$$\mathbb{E}\left[\int_0^t \frac{\alpha_s \mathcal{I}_s}{\sigma_s^2}\, ds\right] = \mathbb{E}\left[\int_0^t \frac{\alpha_s}{\sigma_s^2}(dW(s) - dW'(s))\right]$$

$$= \mathbb{E}\left[\int_0^t \frac{\alpha_s}{\sigma_s^2}\, dW(s)\right] - \mathbb{E}\left[\int_0^t \frac{\alpha_s}{\sigma_s^2}\, dW'(s)\right] = 0;$$

the expectation of each of these two stochastic integrals vanishes, because α_s/σ_s^2 is $\mathscr{F}_s$-measurable and therefore a fortiori $\mathscr{G}_s$-measurable. $\square$

6.5 An Example of an Insider Getting Free Lunches

We assume that the price of the risky asset is given by the SDE

$$dS_t = S_t\big(\sigma(S_t)\, dW_t + \alpha(S_t)\, dt\big) \tag{6.23}$$

where the coefficients σ and α are differentiable functions of S. The insider knows the exact value of the random variable

$$G = \sup_{t \in [0,1]} S_t. \tag{6.24}$$

A practical example of this situation is the exchange rate between two national currencies where some secret bounds are established by the central banks; these bounds being enforced by a direct intervention of central banks on the market. A massive action of central banks on the market could make it necessary to add to (6.23) a local time describing the boundary effect. This discussion can be extended to the possible feedback effect on the stochastic model (6.23) of the insider trading. We shall suppose that the insider trading is done at such a small scale that the stochastic model (6.23) is not violated.

Theorem 6.11. *The insider model defined by* (6.23), (6.24) *satisfies* (6.17), (6.11), (6.12). *Therefore the information drift* $\mathcal{I}$ *exists and satisfies*

$$\mathbb{E}\left[\int_0^1 |\mathcal{I}_s|\, ds\right] < \infty . \tag{6.25}$$

Each $(\mathcal{F}_s)$-*semimartingale is a* $(\mathcal{G}_s)$-*semimartingale; furthermore the infor-mation drift satisfies*

$$\mathbb{E}\left[\int_0^1 |\mathcal{I}_s|^2\, ds\right] = \infty ; \tag{6.26}$$

hence with positive probability the insider has arbitrage opportunities.

Proof. The reader who wants a full proof, going further than the sketch below, should consult Imkeller–Pontier–Weisz [100] and Imkeller [99].

Let $G_t = \sup_{\tau \in [0,t]} S_\tau$; then (G_t, S_t) is a Markov process. Denote by $q_{t,S}(y)\, dy$ the law of $\sup_{s \in [t,1]} S_s$ under the conditioning $S_t = S$. The con-ditional law process satisfies

$$\langle \phi, \mu_t \rangle = \mathbb{E}^{\mathcal{F}_t}[\phi(G)]$$
$$= \phi(G_t) \int_{S_t}^{G_t} q_{t,S_t}(y)\, dy + \int_{G_t}^{\infty} \phi(y) q_{t,S_t}(y)\, dy. \tag{6.27}$$

By the hypothesis of differentiability on the coefficients of SDE (6.24) we have $S_t \in D_1^2$ and also $G_t \in D_1^2$ (see Nualart–Vives [164] and Nualart [159], p. 88). The maximum G_t is reached at a unique point $\tau_t \le t$ and we have $\tau_t < t$ with probability 1; consequently

$$D_t(G_t) = 0 . \tag{6.28}$$

Differentiating (6.27) and using the equation $q_{t,S}(S) = 0$, we get

$$\langle \phi, D_t \mu_t \rangle = (D_t S_t)\left(\phi(G_t) \int_{S_t}^{G_t} \frac{\partial q}{\partial S}(y)\, dy + \int_{G_t}^{\infty} \frac{\partial q}{\partial S}(y)\phi(y)\, dy\right) .$$

Then condition (6.17) is satisfied; condition (6.11) is satisfied as well where the Radon–Nikodym derivative is

$$\delta_t = (D_t S_t)\left(1_{[\tau_1,1]}(t) \frac{\partial}{\partial S} \log\left[\int_S^G q_{t,S}(y)\, dy\right] + \frac{\partial}{\partial S} \log q_{t,S}(G)\right) .$$

From the last expression one can derive (6.25) and (6.26). Condition (6.26) implies that the Girsanov transformation involved to realize the disappear-ance of the drift $\alpha + \mathcal{I}$ diverges; therefore by the Delbaen–Schachermayer theorem [65] the possibility of arbitrage opportunities is open. $\square$

7

Asymptotic Expansion and Weak Convergence

In undergraduate calculus courses the Taylor formula provides the key tool for computing asymptotic expansions. The Stochastic Calculus of Variations started very quickly to play an analogous role in the probabilistic setting, and stochastic Taylor formulae appeared in [2, 10, 23, 25, 26, 38, 50, 126, 127, 139, 176]. In all these developments, the result of Watanabe [213], which provides the methodology of projecting an asymptotic expansion through a non-degenerated map, plays a key role.

A second stream of papers appeared with the application of these methodologies to problems in asymptotic statistics, starting with [214–217] and followed by many others. A third stream finally is concerned directly with mathematical finance where the following papers can be listed: [15–18, 78, 111, 114, 117–123, 128, 137, 138, 153, 154, 177, 198–200, 202, 203]. A reasonable survey of the literature in this direction would go far beyond the limits of this book.

Based on the theory of stochastic flows, the first section of this chapter develops an asymptotic formula for solutions of an SDE depending smoothly on a parameter. The second section presents the theory of Watanabe distributions on Wiener space and its consequences for asymptotic expansions. In particular, we shall deduce an asymptotic expansion for European digital options in a market driven by a uniformly hypoelliptic SDE depending smoothly on a parameter.

The two last sections deal with specific problems of convergence related to the Euler scheme. Strong convergence of the Euler scheme in terms of Sobolev norms on the Wiener space is treated in Sect. 7.3. The fourth section finally is concerned with weak convergence of the scheme in a new space of distributions on the Wiener space, the so-called *diagonal distributions*. The results of the last two sections have been summarized in [146].

7.1 Asymptotic Expansion of SDEs Depending on a Parameter

We consider an $\mathbb{R}^d$-valued SDE depending upon a parameter. Using the notation of (2.2), we write in Stratonovich notation:

$$dS_W^\varepsilon = \sum_{k=1}^n A_k(S_W^\varepsilon(t), \varepsilon) \circ dW^k + A_0(S_W^\varepsilon(t), \varepsilon)\, dt, \quad S_W^\varepsilon(0) = 0. \quad (7.1)$$

Theorem 7.1. *Assume that the vector fields A_k have bounded derivatives of any order in the variables (x, ε), then for any q there exist processes $v_{j,W}(t)$, computable by solving Stratonovich SDEs, such that*

$$\left\| S_W^\varepsilon(t) - S_W^0(t) - \sum_{j=1}^q \varepsilon^j\, v_{j,W}(t) \right\|_{D_r^p(\mathscr{W})} = o(\varepsilon^q), \quad \forall p, r < \infty. \quad (7.2)$$

Proof. In order to shorten the proof we limit ourselves to the case $q = 1$. We extend the SDE to $\mathbb{R}^{d+1}$ by adding an additional equation to system (7.1):

$$d\varepsilon(t) = 0, \quad dS_W^\varepsilon = \sum_{k=1}^n A_k(S_W^\varepsilon(t), \varepsilon) \circ dW^k + A_0(S_W^\varepsilon(t), \varepsilon)\, dt,$$

$$\varepsilon(0) = \varepsilon, \quad S_W^\varepsilon(0) = 0. \quad (7.3)$$

We denote $U_{t \leftarrow 0}^W$ the stochastic flow of diffeomorphisms on $\mathbb{R}^{d+1}$ associated to SDE (7.3) and consider the corresponding tangent flow $J_{t \leftarrow 0}^W$ which has been defined in Chap. 2, (2.13). We remark that the hyperplanes $\{\varepsilon = \text{const}\}$ are preserved by the flow $U_{t \leftarrow 0}^W$. Denote by $e_0, e_1, \ldots, e_d$ the canonical basis of $\mathbb{R}^{d+1}$, the vector e_0 corresponding to the component in ε. Then $(J_{t \leftarrow 0}^W(se_0))(e_0) = (e_0, u_W(s, t))$.

By the elementary integral Taylor formula,

$$S_W^\varepsilon(t) - S_W^0(t) = \varepsilon\, u_W(0, t) + \int_0^\varepsilon (u_W(s, t) - u_W(0, t))\, ds.$$

Taking $v_{1,W}(t) := u_W(0, t)$, the remainder term ρ takes the form

$$\rho := \varepsilon \int_0^1 (u_W(s\varepsilon, t) - u_W(0, t))\, ds = \varepsilon^2 \int\!\!\int_{0 < \tau < s < 1} (\mathcal{J}_{t \leftarrow 0}^W(\tau e_0))(e_0, e_0)\, d\tau ds$$

where $\mathcal{J}_{t \leftarrow 0}$ denotes the flow associated to the second prolongation of SDE (7.3) defined in Sect. 2.6. Now use the fact that $\Phi: W \mapsto \mathcal{J}_{t \leftarrow 0}^W(\tau e_0)$ satisfies bounds of the norms $D_1^p(\mathscr{W})$ uniformly in τ. $\quad \square$

Remark 7.2. Suppose that the market described by SDE (7.1) a digital European option with payoff $f(\varepsilon) := \mathbb{E}[1_K(S)]$ is given. Then it is not clear whether the results of this section imply the existence of an asymptotic expansion for f. However the Watanabe theory will permit us to establish such an expansion.

7.2 Watanabe Distributions and Descent Principle

We defined in Chap. 4, (4.12), the space

$$D^\infty(\mathscr{W}) = \bigcap_{p,r<\infty} D_r^p(\mathscr{W}) \ .$$

A distance on $D^\infty(\mathscr{W})$ is given by the formula

$$\delta(f_1, f_2) = \sum_{p,r=1}^{+\infty} \eta_{p,r}\, 1 \wedge \|f_1 - f_2\|_{D_r^p(\mathscr{W})} \ ,$$

where $\eta_{p,r} > 0$ such that $\sum_{p,r} \eta_{p,r} < \infty$. This topology on $D^\infty(\mathscr{W})$ is independent of the choice of the sequence $\{\eta_{p,r}\}$.

With respect to this distance, $D^\infty(\mathscr{W})$ becomes a complete metric space. We call $D^{-\infty}(\mathscr{W})$ the space of continuous linear forms on $D^\infty(\mathscr{W})$. It results from [66] that, given $T \in D^{-\infty}(\mathscr{W})$, there exist c, p, r such that

$$|\langle f, T \rangle| \le c\, \|f\|_{D_r^p(\mathscr{W})} \ .$$

We define a map $\chi_{\mathscr{W}} \colon D^\infty(\mathscr{W}) \mapsto D^{-\infty}(\mathscr{W})$ by associating to the function g the linear form T_g given by

$$\langle f, T_g \rangle := \mathbb{E}[fg] \ .$$

On $\mathbb{R}^d$ we consider the Schwartz space $\mathscr{S}(\mathbb{R}^d)$ of functions rapidly decreasing together with all their derivatives, and denote by $\mathscr{S}'(\mathbb{R}^d)$ its dual, the space of tempered distributions. We have a similar identification on $\chi_{\mathbb{R}^d} \colon \mathscr{S}(\mathbb{R}^d) \to \mathscr{S}'(\mathbb{R}^d)$ defined by

$$\langle u, T_v \rangle := \int_{\mathbb{R}^d} u(x)v(x)\, dx \ ,$$

where dx is the Lebesgue measure on $\mathbb{R}^d$. Recall that $\chi_{\mathbb{R}^d}(\mathscr{S})$ is dense in $\mathscr{S}(\mathbb{R}^d)$.

Given a non-degenerate map $F \colon \mathscr{W} \mapsto \mathbb{R}^d$ and a function $u \in \mathscr{S}(\mathbb{R}^d)$, we consider the inverse image $F^*u := u \circ F = \tilde{u}$. The fact that $F \in D^\infty(\mathscr{W}; \mathbb{R}^d)$ implies that

$$F^*[\mathscr{S}(\mathbb{R}^d)] \subset D^\infty(\mathscr{W}) \ .$$

Let p be the density of the law of F with respect to the Lebesgue measure. By the results of Chap. 4 we get $p \in \mathscr{S}(\mathbb{R}^d)$ and

$$\int_{\mathbb{R}^d} uv\, p\, d\xi = \mathbb{E}[\tilde{u}\tilde{v}] \ . \tag{7.4}$$

Theorem 7.3 (Watanabe's continuity theorem). *Let F be a non-degenerate map. Given $T \in \mathscr{S}'(\mathbb{R}^d)$, let $u_n \in \mathscr{S}(\mathbb{R}^d)$ be such that $\chi_{\mathbb{R}^d}(u_n)$ converges to T in $\mathscr{S}'(\mathbb{R}^d)$. Then $F^*(u_n)$ converges to $S_T \in D^{-\infty}(\mathscr{W})$ in the topology of $D^{-\infty}(\mathscr{W})$, and the following duality formula holds:*

$$\langle f, S_T \rangle = \langle p \, \mathbb{E}^F[f], T \rangle , \qquad (7.5)$$

where $\mathbb{E}^F$ denotes the conditional expectation with respect to F.

Proof. It's a well-known fact that $\mathscr{S}(\mathbb{R}^d)$ is a topological algebra, and therefore $\mathscr{S}'(\mathbb{R}^d)$ a topological module on $\mathscr{S}(\mathbb{R}^d)$; this implies that $\chi_{\mathbb{R}^d}(p \, u_n)$ converges in $\mathscr{S}'(\mathbb{R}^d)$ to pT. This convergence lifts up to $D^{-\infty}(\mathscr{W})$ by F^* which proves (7.5). $\square$

Theorem 7.4. *Consider a family of maps $F^\varepsilon \in D^\infty(\mathscr{W}; \mathbb{R}^d)$ such that F^ε has an asymptotic expansion in $D^\infty(\mathscr{W})$ up to order q:*

$$\lim_{\varepsilon \to 0} \varepsilon^{-q} \left\| F^\varepsilon - \sum_{j=0}^{q} \varepsilon^j F_j \right\|_{D_r^p(\mathscr{W})} = 0, \quad \forall p, r < \infty .$$

Let $\sigma_{\overrightarrow{W}}(\varepsilon)$ be the Malliavin forward covariance matrix of F^ε, as defined in (5.6), and denote by $\lambda_{\to}^W(\varepsilon)$ its smallest eigenvalue. Assume that

$$\sup_\varepsilon \mathbb{E}[\lambda_{\to}^W(\varepsilon)^{-N}] < \infty, \quad \forall N .$$

Then the law of F^ε has a C^∞-density p^ε with respect to the Lebesgue measure. Furthermore, there exist C^∞-functions f_j on $\mathbb{R}^d$ such that for any $\xi \in \mathbb{R}^d$ with $p^0(\xi) \neq 0$, we have:

$$\lim_{\varepsilon \to 0} \varepsilon^{-q} \left| p^\varepsilon(\xi) - \sum_{j=1}^{q} \varepsilon^j f_j(\xi) \right| = 0 .$$

Proof. See Watanabe [213]. $\square$

7.3 Strong Functional Convergence of the Euler Scheme

Strong convergence of the Euler scheme is a classical fact. We shall refine this result by expressing the convergence upstairs, i.e., on the probability space itself. Convergence there is in terms of Sobolev norms on the Wiener space. This functional convergence could be useful for evaluating the error in Monte-Carlo simulations for look-back options.

We shall deal with the $\mathbb{R}^d$-valued SDE

$$d\xi_W(t) = \sum_{k=1}^{n} A_k(\xi_W(t)) \, dW^k + A_0(\xi_W(t)) \, dt; \quad \xi_W(t^0) = \xi_0 , \qquad (7.6)$$

where W is an n-dimensional Brownian motion on Wiener space $\mathscr{W}^n$ and A_k are smooth bounded vector fields on $\mathbb{R}^d$ with bounded derivatives of any order. Let

$$\mathscr{L} = \frac{1}{2} \sum_{k,\alpha,\beta} A_k^\alpha A_k^\beta D_\alpha D_\beta + \sum_\alpha A_0^\alpha D_\alpha, \quad D_\alpha = \partial/\partial\xi^\alpha,$$

be the associated differential generator.

Given $\varepsilon > 0$, the Euler scheme of mesh ε is defined by the following recursion formula:

$$\xi_{W_\varepsilon}(q\varepsilon) - \xi_{W_\varepsilon}((q-1)\varepsilon) = \sum_{k=1}^n A_k(\xi_{W_\varepsilon}((q-1)\varepsilon)) \left[W^k(q\varepsilon) - W^k((q-1)\varepsilon) \right]$$

$$+ A_0(\xi_{W_\varepsilon}((q-1)\varepsilon))\,\varepsilon, \quad \xi_{W_\varepsilon}(t^0) = \xi_0. \qquad (7.7)$$

Using the notation $t_\varepsilon = t^0 + \left[(t - t^0)/\varepsilon\right]\varepsilon$ where $[a]$ is the largest integer $\leq a$, the Euler scheme for all times is the process defined for $t \in [t^0, T]$ as the solution of the *delayed* SDE

$$d\xi_{W_\varepsilon}(t) = \sum_{k=1}^n A_k(\xi_{W_\varepsilon}(t_\varepsilon))\,dW^k(t) + A_0(\xi_{W_\varepsilon}(t_\varepsilon))\,dt, \quad \xi_{W_\varepsilon}(t^0) = \xi_0.$$

The remainder term $\theta_\varepsilon(t) := \xi_{W_\varepsilon}(t) - \xi_W(t)$ satisfies the SDE

$$d\theta_\varepsilon = \sum_{k=1}^n \left[A_k\big(\xi_W(t) + \theta_\varepsilon(t)\big) - A_k(\xi_W(t)) \right] dW^k(t)$$

$$+ \left[A_0\big(\xi_W(t) + \theta_\varepsilon(t)\big) - A_0(\xi_W(t)) \right] dt + d\chi(t),$$

where $\theta_\varepsilon(t^0) = 0$ and

$$d\chi(t) := \sum_{k=1}^n {}^\varepsilon\mathcal{R}_k(t)\,dW_k(t) + {}^\varepsilon\mathcal{R}_0(t)\,dt,$$

$$^\varepsilon\mathcal{R}_k(t) := A_k(\xi_{W_\varepsilon}(t_\varepsilon)) - A_k(\xi_{W_\varepsilon}(t)). \qquad (7.8)$$

Let $\mathbf{A}_k$ be the $d \times d$ matrix defined by differentiating the components of the vector field A_k with respect to the coordinate vector fields. Then, almost surely, the derivative of the solution $\xi_{W_\varepsilon}(t)$ to (7.7) with respect to the initial data ξ_0 defines a random flow of diffeomorphisms; its Jacobian is given by the matrix-valued delayed SDE

$$d_t J_{t \leftarrow t^0}^{W_\varepsilon} = J_{t \leftarrow t^0}^{W_\varepsilon} \left(\sum_{k=1}^n \mathbf{A}_k(\xi_W(t_\varepsilon))\,dW^k + \mathbf{A}_0(\xi_W(t_\varepsilon))\,dt \right), \quad t \geq t^0,$$

where $J_{t^0 \leftarrow t^0}^{W_\varepsilon} = $ identity.

The derivatives of $^\varepsilon\mathcal{R}_k(t)$ may be computed in terms of the Jacobian matrix:

$$D_{\tau,\ell}\,^\varepsilon\mathcal{R}_k(t) = 1_{\{\tau\le t\}}\left(\mathbf{A}_k(\xi_{W_\varepsilon}(t_\varepsilon))\,J^{W_\varepsilon}_{t\leftarrow\tau}(A_\ell) - \mathbf{A}_k(\xi_{W_\varepsilon}(t))\,J^{W_\varepsilon}_{t\leftarrow\tau}(A_\ell)\right)\;;$$
(7.9)

the derivatives $u(t) := D_{\tau,\ell}\,\theta_\varepsilon(t)$ are computed by differentiating (7.8). We get

$$du - \sum_{k=1}^{n}\mathbf{A}_k(\xi_W + \theta_\varepsilon)\,u\,dW^k - \mathbf{A}_0(\xi_W + \theta_\varepsilon)\,u\,dt =: d\Gamma$$

$$\equiv \sum_{k=1}^{n}\Gamma_k\,dW^k + \Gamma_0\,dt,\qquad(7.10)$$

where $\Gamma_0, \Gamma_1, \ldots, \Gamma_n$ can be computed using (7.9) and standard computations of derivatives along the stochastic flow to SDE (7.6). By Itô's formula, a version of the Lagrange formula (variation of constants) may be established for $u(t)$. To take care of the Itô contraction, the Lagrange formula for ODEs needs to be modified by adding the *compensation vector field* given by

$$Z := \sum_{k=1}^{n}\mathbf{A}_k\,\Gamma_k\,.$$

Then, by Itô's formula, we get

$$u(t) = J_{t\leftarrow t^0}\left[\int_{t^0}^{t}J_{t^0\leftarrow\tau}\left(d\Gamma(\tau) - Z(\tau)\,d\tau\right)\right],\qquad(7.11)$$

where the Itô stochastic integral inside the brackets has to be computed first. We introduce a parameter $\lambda \in [0,1]$ and define $^\lambda\theta$ as the solution of the SDE

$$d^\lambda\theta(t) = \sum_{k=1}^{n}\left[A_k\left(\xi_W(t) + {}^\lambda\theta(t)\right) - A_k(\xi_W(t))\right]dW^k(t)\qquad(7.12)$$

$$+ \left[A_0\left(\xi_W(t) + {}^\lambda\theta(t)\right) - A_0(\xi_W(t))\right]dt + \lambda\,d\chi(t),\quad {}^\lambda\theta(t^0) = 0.$$

As $^0\theta(t) = 0$ for all t, denoting $\frac{d}{d\lambda}{}^\lambda\theta = {}^\lambda u$, we have $\theta_\varepsilon = \int_0^1 {}^\lambda u\,d\lambda$. By differentiating (7.12) with respect to λ, we get the following linear SDE for $^\lambda u$:

$$d^\lambda u = d^\varepsilon\mathcal{Q}\cdot{}^\lambda u + d\chi\,,\qquad(7.13)$$

where $d^\varepsilon\mathcal{Q} = \sum_{k=1}^{n}\mathbf{A}_k(\xi_W + {}^\lambda\theta)\,dW^k + \mathbf{A}_0(\xi_W + {}^\lambda\theta)\,dt$.

Theorem 7.5. *For any $p \in \,]1,\infty[$ and any integer $r > 0$, we have*

$$\left(\mathbb{E}\left[\sup_{t\in[t^0,T]}|\theta_\varepsilon(t)|_{\mathbb{R}^d}^p\right]\right)^{1/p} = O(\sqrt{\varepsilon}); \qquad \sup_{t\in[t^0,T]}\left\|\theta_\varepsilon(t)\right\|_{D_r^p} = O(\sqrt{\varepsilon})\,.\qquad(7.14)$$

Proof. Denote by ${}^{\varepsilon}J^W_{t\leftarrow\tau}$ the solution to the linear equation (7.13) with initial condition ${}^{\varepsilon}J^W_{\tau\leftarrow\tau} = \mathrm{id}$; as its coefficients $\mathbf{A_k}$ are bounded, we have uniformly with respect to ε:

$$\mathbb{E}\left[\sup_{t,\tau\in[t^0,T]}\left|{}^{\varepsilon}J^W_{t\leftarrow\tau}\right|^{2p}\right] \le c_p < \infty .$$

For the same reason the compensation vector field Z is bounded in L^p; we get $\mathbb{E}\left[|{}^{\varepsilon}\mathcal{R}_k|^{2p}_{\mathbb{R}^d}\right] = O(\varepsilon^p)$. Using (7.11), we have

$$\lambda_u(t) = {}^{\varepsilon}J^W_{t\leftarrow t^0}\left[\int_{t^0}^t \sum_{k=1}^n {}^{\varepsilon}J^W_{t^0\leftarrow\tau}\,{}^{\varepsilon}\mathcal{R}_k(\tau)\,dW^k(\tau) + {}^{\varepsilon}J^W_{t^0\leftarrow\tau}\left({}^{\varepsilon}\mathcal{R}_0(\tau) - Z(\tau)\right)d\tau\right] .$$

Consequently $D_{\tau,k}\theta$ satisfies an SDE of the same nature as (7.8); the second part of (7.14) is verified along the same lines. $\square$

7.4 Weak Convergence of the Euler Scheme

The "weakest" among weak topologies are topologies on spaces of distributions; we adopt this point of view in this section. Our main concern is to establish asymptotic expansions "upstairs", i.e., on the probability space itself. The Watanabe principle then makes it possible to pass from upstairs estimates to estimates of densities.

We recall that the norm of the rth derivatives of f has been defined as

$$\left(\int_0^1 \cdots \int_0^1 |D_{\tau_1,\dots,\tau_r}f|^2\,d\tau_1\dots d\tau_r\right)^{1/2} .$$

The Sobolev spaces D^p_r are defined with respect to this norm.

Consider the 1-dimensional subspace of $\mathbb{R}^s$ given by the diagonal $\{\tau_1 = \dots = \tau_s\}$ and define a function V^s on $\mathbb{R}^s$ by $V^s(\tau_1,\dots,\tau_s) = D_{\tau_1,\dots,\tau_s}f$ if $\tau_i \in [0,1]$ and $V^s = 0$ otherwise. We make an orthonormal change of coordinates $(\eta_1,\dots,\eta_s)$ such that the diagonal is given by the equation $\zeta = 0$ where $\zeta = (\eta_2,\dots,\eta_s)$. Denoting by $W^s(\eta_1,\zeta)$ the function V^s in this new system of coordinates, we consider the partial functions $W^s_\zeta\colon \eta_1 \mapsto W^s(\eta_1,\zeta)$. The function V^s is said to be *diagonal-continuous* if it has a version such that $\zeta \mapsto W^s_\zeta$ is a continuous map from $\mathbb{R}^{r-1}$ to $L^2(\mathbb{R})$. We denote by ${}^\gamma D^p_r$ the subspace of D^p_r such that the second up to the rth derivative is diagonal-continuous. A diagonal Sobolev norm is then defined by

$$\|f\|^p_{{}^\gamma D^p_r} = \mathbb{E}\left[|f|^p + \left(\int_0^1 (D_\tau f)^2\,d\tau\right)^{p/2} + \sum_{s=2}^r \sup_{\zeta\in\mathbb{R}^{s-1}}\|W^s_\zeta\|^p_{L^2(\mathbb{R})}\right] .$$

Definition 7.6. *A diagonal distribution (see [146]) is a linear form S on $D_{r\gamma}^p$ such that*

$$|\langle f, S \rangle| \le c \, \|f\|_{\gamma D_r^p} \, .$$

We localize the method of Sect. 7.3 by using a Taylor expansion at $\lambda = 0$. Denoting the second derivatives by $^\lambda v := (d/d\lambda)\,^\lambda u$ and using Theorem 7.5, we have

$$^\lambda\theta = \lambda \, {}^\circ u + \frac{\lambda^2}{2} \, {}^\circ v + o(\varepsilon)$$

with $o(\varepsilon)$ being uniform in λ. The question of asymptotic expansion of θ as $o(\varepsilon)$ is therefore reduced to the asymptotic expansion of ${}^\circ u, {}^\circ v$ which we abbreviate as u, v. We compute u from (7.13) for $\lambda = 0$ which realizes a computation along the path of the original diffusion. In the same way, if (7.13) is differentiated with respect to λ and if λ is set equal to 0, the terms

$$\mathcal{R}_k^1(t) := \sum_{\ell,p=1}^d \frac{\partial^2 A_k}{\partial \xi^\ell \partial \xi^p}(\xi_W(t))\, u^\ell(t) u^p(t), \quad Z^1(t) := \sum_{k=1}^n (\mathbf{A}_k \cdot \mathcal{R}_k^1)\,(\xi_W(t))$$

(7.15)

appear, and we get

$$v(t) = J_{t \leftarrow t^0}^W \left[\int_{t^0}^t \sum_{k=1}^n J_{t^0 \leftarrow \tau}^W (\mathcal{R}_k^1(\tau))\, dW^k(\tau) \right.$$

$$\left. + J_{t^0 \leftarrow \tau}^W \left(\mathcal{R}_0^1(\tau) - Z^1(\tau) \right) d\tau \right]. \qquad (7.16)$$

Theorem 7.7 (Rate of weak convergence). *There exist $\mathbb{R}^d$-valued functions $a_{k\ell}, b_k, c$ on $\mathbb{R}^d$, computable in terms of the coefficients of (7.6) and its first fourth derivatives, such that for any $f \in {}^\gamma D_3^{\infty-0}$,*

$$\lim_{\varepsilon \to 0} \frac{1}{\varepsilon} \mathbb{E}[\theta_\varepsilon(t)\, f] = \int_{t^0}^t \mathbb{E} \left[\sum_{k,\ell}^n a_{k\ell}(\xi_W(\tau))\, D^2_{\tau,k;\tau,\ell} f \right.$$

$$\left. + \sum_{k=1}^n b_k(\xi_W(\tau))\, D_{\tau,k} f + c(\xi_W(\tau))\, f \right] d\tau \, .$$

Proof. Assume that diagonal distributions $\tilde{S}^1, \dots, \tilde{S}^d$ are given and consider the formal expression $S^q = \sum_k (J_{t \leftarrow t^0}^W)_k^q \, \tilde{S}^k$. As the coefficients of the matrix $J_{t^0 \leftarrow t}^W$ belong to ${}^\gamma D_3^{\infty-0}$ and as the space of diagonal distributions is a module over the algebra ${}^\gamma D_3^{\infty-0}$, we deduce that S is a diagonal distribution. Defining

$$\tilde{u}(t) := J_{t^0 \leftarrow t}(u(t)), \quad \tilde{v}(t) := J_{t^0 \leftarrow t}(v(t)) \, ,$$

we are reduced to finding a diagonal distribution $\tilde{S}$ such that

$$\lim_{\varepsilon \to 0} \frac{1}{\varepsilon} \mathbb{E}[\tilde{u}(t) f] = \langle f, \tilde{S} \rangle, \quad \lim_{\varepsilon \to 0} \frac{1}{\varepsilon} \mathbb{E}[\tilde{v}(t) f] = 0 \, .$$

To short-hand the notation, we write $^W(\cdot) := J^W_{t^0 \leftarrow \tau}(\cdot)$; then $\tilde{u}(t) = \tilde{u}_1(t) + \tilde{u}_2(t)$ where

$$\tilde{u}_1(t) := \int_{t^0}^t \sum_{k=1}^n {}^W(^\varepsilon\mathcal{R}_k(\tau))\, dW^k(\tau), \quad \tilde{u}_2(t) := \int_{t^0}^t {}^W(^\varepsilon\mathcal{R}_0(\tau) - Z(\tau))\, d\tau .$$

Using the fact that an Itô integral is a divergence we get

$$\mathbb{E}[f\tilde{u}_1(t)] = \int_{t^0}^t \mathbb{E}\left[\sum_{k=1}^n (D_{\tau,k}f)\, {}^W(^\varepsilon\mathcal{R}_k(\tau))\right] d\tau .$$

Since $^\varepsilon\mathcal{R}_k(\tau_\varepsilon) = 0$, we get by applying Itô's formula,

$$^\varepsilon\mathcal{R}_k(\tau) = -\int_{\tau_\varepsilon}^\tau \sum_s \mathbf{A}_k(\xi_{W_\varepsilon}(\lambda)) \cdot A_s(\xi_{W_\varepsilon}(\tau_\varepsilon))\, dW^s(\lambda) - \int_{\tau_\varepsilon}^\tau (^\varepsilon\mathscr{L}A_k)\,(\xi_{W_\varepsilon}(\lambda))\, d\lambda$$

$$(7.17)$$

where

$$(^\varepsilon\mathscr{L}A_k)(\xi_{W_\varepsilon}(\lambda)) := \frac{1}{2} \sum_{s,\ell,p} A_s^\ell(\xi_{W_\varepsilon}(\tau_\varepsilon))\, A_s^p(\xi_{W_\varepsilon}(\tau_\varepsilon)) \frac{\partial^2 A_k}{\partial \xi_\ell \partial \xi_p}(\xi_{W_\varepsilon}(\lambda))$$

$$+ \sum_\ell A_0^\ell(\xi_{W_\varepsilon}(\tau_\varepsilon)) \frac{\partial A_k}{\partial \xi_\ell}(\xi_{W_\varepsilon}(\lambda)) .$$

We want to eliminate the stochastic integral in (7.17). For this purpose, we remark

$$\mathbb{E}\left[(D_{\tau,r}f)\, {}^W(^\varepsilon\mathcal{R}_r(\tau))\right] = \mathbb{E}\left[\mathbb{E}^{\mathcal{N}_\tau}(D_{\tau,r}f)\, {}^W(^\varepsilon\mathcal{R}_r(\tau))\right] ;$$

$$\mathbb{E}\left[\mathbb{E}^{\mathcal{N}_{\tau_\varepsilon}}(D_{\tau,r}f) \int_{\tau_\varepsilon}^\tau \sum_s \mathbf{A}_k(\xi_{W_\varepsilon}(\lambda)) \cdot A_s(\xi_{W_\varepsilon}(\tau_\varepsilon))\, dW^s(\lambda)\right] = 0 .$$

From the Ocone–Karatzas formula, we get

$$\mathbb{E}^{\mathcal{N}_\tau}[D_{\tau,r}f] - \mathbb{E}^{\mathcal{N}_{\tau_\varepsilon}}[D_{\tau,r}f] = \sum_{k=1}^n \int_{\tau_\varepsilon}^\tau \mathbb{E}^{\mathcal{N}_\lambda}(D^2_{\tau,r;\lambda,k}f)\, dW^k(\lambda)$$

and with the short-hand notation $^W(\cdot)$ analogous as above,

$$\mathbb{E}[f\tilde{u}_1(t)] = \mathbb{E}\left[\int_{t^0}^t d\tau \int_{\tau_\varepsilon}^\tau d\lambda \left(\sum_{r,k=1}^n (D^2_{\tau,r;\lambda,k}f)\, {}^W\left(\mathbf{A}_r(\xi_{W_\varepsilon}(\lambda)) \cdot A_k(\xi_{W_\varepsilon}(\tau_\varepsilon))\right)\right.\right.$$

$$\left.\left. + \sum_{k=1}^n (D_{\tau,k}f)\, {}^W(^\varepsilon\mathscr{L}A_k)(\xi_{W_\varepsilon}(\lambda))\right)\right]$$

which as $\varepsilon \to 0$ gives rise to the equivalence

$$\mathbb{E}[f\tilde{u}_1(t)] \simeq \frac{\varepsilon}{2}\mathbb{E}\left[\int_{t^0}^t d\tau \left(\sum_{r,k=1}^n (D_{\tau,r;\tau,k}^2 f)\,{}^W((\mathbf{A}_r \cdot A_k)(\xi_W(\tau)))\right.\right.$$

$$\left.\left. + \sum_{k=1}^n (D_{\tau,k}f)\,{}^W({}^0\mathscr{L}A_k)(\xi_W(\tau))\right)\right].$$

Finally

$$\mathbb{E}[f\tilde{u}_2(t)] = \mathbb{E}\left[\int_{t^0}^t f \cdot ({}^W\mathcal{R}_0 - {}^WZ)\,d\tau\right]$$

may be computed as before. Now integrals along the paths of $D_{\tau,k}f$, f appear, and we get coefficients $\hat{a}, \hat{b}, \hat{c}$ such that

$$\lim_{\varepsilon \to 0} \frac{1}{\varepsilon}\mathbb{E}[\tilde{u}(t)f] = \int_{t^0}^t \mathbb{E}\left[\sum_{k,\ell}^n \hat{a}_{k\ell}(\xi_W(\tau))\,D_{\tau,k;\tau,\ell}^2 f\right.$$

$$\left. + \sum_{k=1}^n \hat{b}_k(\xi_W(\tau))\,D_{\tau,k}f + \hat{c}(\xi_W(\tau))\,f\right]d\tau. \qquad (7.18)$$

We are left to deal with $\tilde{v}$: combining (7.14) and (7.16) gives $\tilde{v} = O(\varepsilon)$. To get the sharper estimate $o(\varepsilon)$ we use that (7.15) expresses $\mathcal{R}_k^1$ as a bilinear functional in u. A typical term is

$$\mathbb{E}\left[f\int_{t^0}^t \frac{{}^W\partial^2 A_k}{\partial\xi^\ell\partial\xi^p}\,u^\ell u^p\,dW^k\right] = \int_{t^0}^t \mathbb{E}[\mathcal{B}(\tau)\,u^p(\tau)]\,d\tau\ ,$$

$$\text{where } \mathcal{B} := (D_{\tau,k}f)\,\frac{{}^W\partial^2 A_k}{\partial\xi^\ell\partial\xi^p}\,u^\ell;$$

here we used the Clark–Ocone–Karatzas formula for f. Applying formula (7.18) with $f = \mathcal{B}$, we get

$$\left|\mathbb{E}\left[f\int_{t^0}^t \frac{{}^W\partial^2 A_k}{\partial\xi^\ell\partial\xi^p}\,u^\ell u^p\,dW^k\right]\right| \leq c\varepsilon\,\|\mathcal{B}\|_{\gamma D_2^s} \leq c\varepsilon\,\|f\|_{\gamma D_3^{2s}}\,\|u\|_{\gamma D_2^{2s}}\ ,$$

and using (7.14), with $\theta_\varepsilon(\cdot)$ replaced by u, we obtain the required order of convergence. $\square$

8

Stochastic Calculus of Variations for Markets with Jumps

Stochastic Calculus of Variations for jump process can be developed following three different paradigms.

The paradigm which appeared first in the literature is to make a variation on the intensity of jumps, the time when the jumps occur being fixed (see [22, 36, 155]). This methodology gives in particular deep regularity results for local times. As it has not yet been used in mathematical finance we shall not discuss it here.

The second paradigm uses an approach of chaos expansion parallel to the approach presented in Sect. 1.6 for the Wiener space. A beautiful conceptual theory can then be built upon this approach [31, 161, 162, 179]: this theory is as complete as the theory on the Wiener space. Nevertheless certain objects appearing there are difficult to realize effectively by a trader operating on the market.

The third paradigm is to build the theory on a concept of *pathwise instantaneous derivative*. This approach, perfectly fitting to the computation of Greeks, has a simple conceptual meaning and corresponds to operations which can be easily implemented by a market practitioner; on the other hand it introduces structural incompleteness to the market.

In Sect. 8.1 we construct the probability spaces for finite type jump processes. Section 8.2 develops the Stochastic Calculus of Variations for exponential variables, and Sect. 8.3 the Stochastic Calculus of Variations for Poisson processes. Section 8.4 finally establishes mean-variance minimal hedging through a weak version of a Clark–Ocone formula.

We shall not develop the effective computation of Monte-Carlo weights for the computation of Greeks, as it was done in [70, 180] by using the formula of integration by parts on Poisson spaces.

8.1 Probability Spaces of Finite Type Jump Processes

The general theory of scalar-valued, time-homogeneous, Markov jump processes considers processes which in any time interval may have an infinite number of jumps: the *Lévy processes*.

We call such a process of *finite type* if almost surely on any finite interval the number of jumps is finite. This class of processes is qualitatively sufficient for the needs of mathematical finance; we shall limit our study to this class in order to avoid technical difficulties inherent to the general theory of Lévy processes. Finally it is well known that the class of processes of finite type is dense in the class of all Lévy pro-cesses.

Theorem 8.1 (Structural theorem).   *Let $X(t)$ be a scalar-valued, time-homogeneous Markov process of finite type. Then there exists a Poisson process $N(t)$ of intensity 1, a Brownian motion W, and three constants σ, c, ρ such that*

$$X(t) - X(0) = ct + \sigma W(t) + \sum_{1 \le i \le N(\rho t)} Y_i \,, \tag{8.1}$$

where $(Y_i)_{i \ge 1}$ is a family of independent equi-distributed random variables.

Proof. This is a classical fact.   □

Remark 8.2. An alternative point of view is to take formula (8.1) as the definition of processes of finite type.

We consider jump process defined on the time interval $[0, +\infty[$ and denote by $\mathscr{W}$ the probability space of scalar-valued Brownian motion indexed by this interval. We further denote by μ the law of Y_1 and introduce the probability space Ω^μ associated to a countable sequence of independent random variables equi-distributed according to μ. This space can be realized on the infinite product $\mathbb{R}^{\mathbb{N}}$ equipped with the infinite product measure $\bigotimes_k \mu_k$ where $\mu_k = \mu$.

Consider the special case where μ is the exponential law ν, given by the distribution of an exponential variable T defined as

$$\mathbb{P}\{T > a\} = \exp(-a), \quad a > 0,$$

and denote $\mathcal{N} := \Omega^\nu$ the Poisson probability space.

Theorem 8.3. *The probability space Ω of a finite type jump process is the direct product*

$$\Omega \simeq \mathscr{W} \times \Omega^\mu \times \mathcal{N} \,. \tag{8.2}$$

In the case where the law μ of the random variable Y_1 is supported by q points, we have

$$\Omega \simeq \mathscr{W} \times \mathcal{N}^q \,. \tag{8.3}$$

Remark 8.4. As any law μ can be obtained as weak limit of laws concentrated on a finite number of points, the case (8.3) seems to be sufficient for the need of numerical finance. See [132], p. 215 for a precise scheme of the approximation.

Proof. Our Poisson process on $\mathbb{R}^+$ is uniquely given by an independent sequence $(T_n)_{n \in \mathbb{N}}$ of identically distributed exponential variables:

$$\mathbb{P}\{T_n > a\} = \int_a^{+\infty} \exp(-s)\,ds = \exp(-a) \ .$$

We define the probability space Ω of the Poisson process as the probability space corresponding to this sequence $\{T_k\}$ of independent exponential variables. Denote by $\mathscr{G}_k$ the σ-field generated by the first k coordinates $T_1, \ldots, T_k$. The first jump of the Poisson process appears at T_1, the second at $T_1 + T_2$, and so on. The Poisson counting function is

$$N(t) = \sup\{k : T_1 + \ldots + T_k \le t\} \ . \tag{8.4}$$

We are going to prove (8.3). Let

$$\mu = \sum_{s=1}^q p_s\,\delta_{\xi_s}$$

where δ_ξ denotes the Dirac mass at the point ξ. Taking q independent Poisson processes N_s, then

$$X(t) = \sigma W(t) + \sum_{s=1}^q \xi_s N_s(\rho_s t). \quad \square \tag{8.5}$$

Itô Stochastic Integral on Processes of Finite Type

We recall the definition of càdlàg functions on $\mathbb{R}^+$, constituted by the class of functions $t \mapsto F(t)$ which are right-continuous with limits from the left; for any $t \in \mathbb{R}^+$, $\lim_{0 < \varepsilon \to 0} F(t+\varepsilon) = F(t)$ and $\lim_{0 < \varepsilon \to 0} F(t-\varepsilon)$ exists. Note that the Poisson process as defined by (8.4) has càdlàg paths; by consequence the process $X(t)$ is also càdlàg.

We denote by $(\mathscr{F}_t)$ the filtration generated by $X(\cdot)$:

$$\mathscr{F}_t = \sigma\text{-field generated by } X(s), \quad s \in [0, t] \ .$$

This filtration is right-continuous. We have two basic definitions:

- a process Y is *adapted* if $Y(t)$ is $\mathscr{F}_t$-measurable;
- a process Z is *predictable* if Z is left-continuous and adapted.

In the case where the jump component disappears, the filtration $(\mathscr{F}_t)$ is also left-continuous and the notions of adaptedness and of predictability coincide. This is not the case in general; for instance, the process $X(\cdot)$ is adapted but is not predictable.

The *stochastic integral of a predictable process Z* is defined as limit of the Riemann sums:

$$\int_0^1 Z(s)\,dX(s) = \lim_{p\to\infty} \sum_{k=0}^{2^p-1} Z\left(\frac{k}{2^p}\right)\left[X\left(\frac{k+1}{2^p}\right) - X\left(\frac{k}{2^p}\right)\right]. \qquad (8.6)$$

From this definition it results that the process defined by the indefinite integral

$$M(t) := \int_0^1 Z(s)\,1_{[0,t]}(s)\,dX(s)$$

is càdlàg.

Proposition 8.5 (Energy identity). *If Z a predictable process, then*

$$\mathbb{E}\left[\int_0^1 Z\,dX\right] = (c + \rho\,\mathbb{E}(Y_1))\,\mathbb{E}\left[\int_0^1 Z\,dt\right]. \qquad (8.7)$$

Assume that $c + \rho\,\mathbb{E}(Y_1) = 0$, then $M(t) := \int_0^t Z\,dX$ is a $(\mathscr{F}_t)$-martingale and its L^2-energy is given by

$$\mathbb{E}\left[\left(\int_0^1 Z\,dX\right)^2\right] = (\sigma^2 + \rho\,\mathbb{E}(Y_1^2))\,\mathbb{E}\left[\int_0^1 Z^2\,dt\right]. \qquad (8.8)$$

Proof. This is classical L^2 martingale theory. $\square$

It is clearly not possible for the trader to react on an infinitesimal scale of time, the most general possible *trading strategies* are given by a predictable process. *Replicable assets* on $[0,T]$ are therefore represented by stochastic integrals on $[0,T]$ of a trading strategy.

8.2 Stochastic Calculus of Variations for Exponential Variables

Denote by Ω the probability space corresponding to a sequence of identically distributed exponential variables T_n. In logarithmic scale, let $\tau_n = \log T_n$: we get $\mathbb{P}\{\tau_n > \alpha\} = \exp(-\exp(\alpha))$. For a single exponential variable we have the following formula of integration by parts:

$$\mathbb{E}[\phi'(\tau)] = \mathbb{E}\left[\phi(\tau)\,\delta\left(\frac{d}{d\tau}\right)\right], \quad \delta\left(\frac{d}{d\tau}\right) := \exp(s) - 1. \qquad (8.9)$$

In fact, we have

$$\mathbb{E}\left[\frac{d\phi}{d\tau}(\tau)\right] = \int_{-\infty}^{+\infty} \phi'(s) \, \exp(-\exp(s)) \, \exp(s) \, ds$$

$$= -\int_{-\infty}^{+\infty} \phi(s) \, \exp(-\exp(s)) \, \exp(s)(1 - \exp(s)) \, ds \; .$$

The logarithmic change of coordinates has the drawback that the traditional scale for exponential variables is left. We transfer the derivation operator from the logarithmic scale to the usual scale by introducing a slightly modified derivation:

$$s\left(\frac{d}{ds}\psi\right)(s) =: \left(\frac{d^\dagger}{ds}\psi\right)(s) = (\psi^\dagger)(s) \; .$$

Lemma 8.6 (Integration by parts for a single exponential variable).
We have

$$\mathbb{E}\left[\psi^\dagger(T)\right] = \mathbb{E}\left[\psi(T)\,\delta\left(\frac{d^\dagger}{dT}\right)\right], \quad \delta\left(\frac{d^\dagger}{dT}\right) := 1 - T; \qquad (8.10)$$

$$\mathbb{E}\left[\left(\delta\left(\frac{d^\dagger}{dT}\right)\right)^2\right] = \int_0^\infty (1-s)^2 \exp(-s)\, ds = 1. \qquad (8.11)$$

Proof. See the previous computations. □

Derivation Operator on Ω

Denote by $C_b^1(\Omega)$ the functionals Φ on Ω depending on a finite number of coordinates $T_1, \ldots, T_q$ such that the dependence is one time differentiable, the derivatives being uniformly bounded. Let

$$\tilde{D}_k \Phi := T_k \frac{\partial \Phi}{\partial T_k} \; . \qquad (8.12)$$

Denote by ℓ^2 the sequence $\{w_k(\omega)\}_{k>0}$ such that $\sum_k \mathbb{E}[w_k^2] < \infty$ and such that w_k is $\mathcal{G}_{k-1}$ measurable. Define

$$\tilde{D}_w \Phi := \sum_k w_k \tilde{D}_k \Phi = \sum_k w_k \, T_k \frac{\partial \Phi}{\partial T_k} \; .$$

Theorem 8.7 (Cameron–Martin). *The series*

$$\sum_k w_k \, (1 - T_k) =: \tilde{\delta}(w) \qquad (8.13)$$

converges in $L^2(\Omega)$; we have the formula of integration by parts

$$\mathbb{E}[\tilde{D}_w \Phi] = \mathbb{E}[\Phi \, \tilde{\delta}(w)] \; . \qquad (8.14)$$

Proof. We have $\mathbb{E}[1 - T_k] = 0$; therefore the partial sum of the series (8.13) constitutes a $(\mathscr{G}_k)$-martingale which is L^2-integrable. □

Remark 8.8. Formula (8.12) implies that the operators $D_k\Phi$ are closable. Denote by $D_1^2(\Omega)$ the functions $\Psi \in L^2(\Omega)$ belonging to the intersection of all domains such that

$$\|\Psi\|_{D_1^2}^2 = \mathbb{E}[\Psi^2] + \sum_k \mathbb{E}[(D_k\Psi)^2] < \infty .$$

Then $D_1^2(\Omega)$ is a Hilbert space.

8.3 Stochastic Calculus of Variations for Poisson Processes

On the probability space $\mathcal{N}$ of the Poisson process, we denote by S_1 the time of the first jump, by S_2 the time of the second jump, and so on. We realize an isomorphism $\mathcal{N} \to \Omega$ by the change of variable between $S.$ and $T.$:

$$S_k = \sum_{j=1}^k T_j, \text{ which implies } \frac{\partial\Phi}{\partial T_k} = \sum_{j\geq k} \frac{\partial\Phi}{\partial S_j} \text{ for the derivatives.}$$

The counting function of the Poisson process is $N(t) = k$, if $t \in [T_k, T_{k+1}[$. Therefore the Stieltjes integral of the compensated Poisson process $N(t) - t$ satisfies the relation

$$\int 1_{[S_{k-1}, S_k[} \, d(N(t) - t) = 1 - T_k .$$

Note that $1_{[S_{k-1},S_k[}$ is predictable. To a predictable weight w_k we associate the predictable process

$$m(t) = \sum_k w_k \, 1_{]S_{k-1},S_k]}$$

and consider the following stochastic integral

$$\int_{\mathbb{R}^+} m(t) \, d(N(t) - t) = \sum_k w_k(1 - T_k) .$$

Denote by H^1 the Hilbert space of functions with derivative in L^2 and vanishing at $t = 0$. Let $v\colon \mathcal{N} \to H^1$ be a map such that $\mathbb{E}[\|v\|_{H^1}^2] < \infty$. We define the derivative $D_v\Phi$ by

$$D_v\Phi := \sum_k v(S_k) \frac{\partial\Phi}{\partial S_k} . \tag{8.15}$$

Theorem 8.9. *Let v be a predictable random process such that*

$$\mathbb{E}\big[\|v\|_{H^1}^2\big] < \infty .$$

Then

$$\mathbb{E}\big[D_v F\big] = \mathbb{E}\big[F\delta(v)\big], \quad \delta(v) = \int_0^{+\infty} v'(t)\, d(N(t) - t) . \tag{8.16}$$

Furthermore, we have the energy identity

$$\mathbb{E}\big[|\delta(v)|^2\big] = \mathbb{E}\big[\|v\|_{H^1}^2\big] . \tag{8.17}$$

Proof. Previous computations combined with Sect. 8.3. □

We extend this situation to the simple jump process of the form

$$X(t) = W(t) + \sum_{k=1}^q \xi_k N_k(\rho_k t) \tag{8.18}$$

where W is a Brownian motion and where N_k are Poisson processes, all processes being independent.

Denote by $\{S_r^k\}_r$ the sequence of jumps of the Poisson process N_k. Take as tangent space $[H^1]^{q+1}$. To $v = (v_0, v_1, \ldots, v_q) \in [H^1]^{q+1}$, associate the derivative

$$D_v \Phi = \sum_{k=0}^q \int D_{t,k} \Phi\, \dot{v}_k(t)\, dt \tag{8.19}$$

where $D_{t,0}$ is the derivative with respect to $\mathscr{W}$ (see Chap. 1) which vanishes on the other components; the derivative D_{t,k_0} acts only on the component N_{k_0} and vanishes on the other components, the action on N_{k_0} being given by

$$D_{t,k_0}(S_q^{k_0}) = \rho_{k_0}\, 1_{[0, S_q^{k_0}[}(\rho_{k_0} t) . \tag{8.20}$$

Remark that the r.h.s. is a predictable function of t. As the jump times determine the Poisson process N_k, any functional on N_k can be expressed in terms of the S_q^k.

The following theorem establishes a remarkable parallelism between the Stochastic Calculus of Variations on Wiener spaces and on Poisson spaces.

Theorem 8.10 (Carlen–Pardoux [47]). *Let $v\colon \Omega \to [H^1(\mathbb{R}^+)]^{q+1}$ be a predictable map such that $\mathbb{E}[\|v\|_{H^1}^2] < \infty$. Then the following formula of integration by parts holds:*

$$\mathbb{E}\big[D_v \Phi\big] = \mathbb{E}\left[\Phi \sum_{k=0}^q \int \dot{v}_k\, dX^k\right] , \tag{8.21}$$

where $dX^0(t) := dW(t)$ and $dX^k := \rho_k(dN_k(t) - dt)$ is the compensated Poisson process.

Proof. Apply (8.16). □

Compare [70, 180] for a computation of Greeks through the formula of integration by parts.

As consequence of Theorem 8.10, the derivative D_v is a closable operator. We denote by $D_1^2(\mathcal{N})$ the Sobolev space of functionals on $\mathcal{N}$ whose first derivative is in L^2.

Remark 8.11. Fixing $t_0 > 0$, consider the random variable $\Psi = N(t_0)$; then $\Psi \notin D_1^2(\mathcal{N})$. In fact, if we express $N(t_0)$ in terms of the S_k by $N(t_0) = \sum_{k=1}^\infty 1_{[0,t_0]}(S_k)$ and if we apply the chain rule, we have to deal with the difficulty that the indicator function of $[0, t_0]$ is not differentiable in L^2. This difficulty can be overcome by developing a theory of distributions on Poisson space [1, 67]. The goal of the next proposition is to give examples of differentiable averages of $N(t)$.

Proposition 8.12. *Let $F(t, k, \omega)$ be a function such that $F(t, k) \equiv F(t, k, \cdot) \in D_1^2(\mathcal{N})$ and denote $(\nabla F)(t, k) = F(t, k+1) - F(t, k)$. Then*

$$D_v \int_0^\infty F(t, N(t))\, dt = -\int_0^\infty v_t\, (\nabla F)(t, N_-(t))\, dN(t) + \int_0^\infty [D_v F](t, N(t))\, dt ,$$

(8.22)

where $N_-(t) = \lim_{t' < t,\, t' \to t} N(t')$.

Proof. Express the integral in terms of the sequence of jumping times S_k of $N(t)$:

$$\int_0^\infty F(t, N(t))\, dt = \sum_{k=1}^\infty \int_{S_{k-1}}^{S_k} F(t, k-1)\, dt .$$

The derivative with respect to ω in the integral gives rise to the second term of (8.22); it remains to differentiate the S_k. As $D_v(S_k) = v(S_k)$, and using the fact that $v(0) = 0$, we get

$$\sum_{k=1}^\infty F(S_k, k-1)\, v(S_k) - F(S_{k-1}, k-1)\, v(S_{k-1})$$

$$= \sum_{k=1}^\infty v(S_k)\, (F(S_k, k-1) - F(S_k, k))$$

$$= -\sum_k v(S_k)\, (\nabla F)(S_k, N_-(S_k))$$

$$= -\int_0^\infty v(t)\, (\nabla F)(t, N_-(t))\, dN(t). \quad \square$$

8.4 Mean-Variance Minimal Hedging and Clark–Ocone Formula

Consider a market for which the underlying probability space takes the form (8.18). To simplify the notations, let $\sigma = 1$, $\xi_1 = 1$, $q = 1$. Denote by $\mathscr{P}$

the predictable processes $w = (w_0, w_1)$. Given $F \in D_1^2(\mathcal{W} \times \mathcal{N})$ which is $\mathcal{G}_1$-measurable such that $\mathbb{E}[F] = 0$, *mean-variance minimal hedging* consists in finding $v \in \mathcal{P}$ such that

$$
\mathbb{E}\left[\left| F - \int_0^1 \dot{v}_0\, dW + \dot{v}_1\, d(N - t) \right|^2 \right]
$$

$$
= \inf_{w \in \mathcal{P}} \mathbb{E}\left[\left| F - \int_0^1 \dot{w}_0\, dW + \dot{w}_1\, d(N - t) \right|^2 \right].
$$

Theorem 8.13. *The mean-variance minimal hedging problem has a unique solution given by*

$$
\dot{v}_0(t) = \mathbb{E}^{\mathcal{G}_t^-}\left[D_t^0 F \right], \quad \dot{v}_1(t) = \mathbb{E}^{\mathcal{G}_t^-}\left[D_t^1 F \right], \tag{8.23}
$$

where $\mathcal{G}_t^- := \sigma(\bigcup_{t' < t} \mathcal{G}_{t'})$ is the predictable filtration.

Proof. Let V be the vector space of random variables representable as stochastic integrals of predictable processes. It results from the energy identity (8.8) that V forms a closed subspace of $L^2(\mathcal{W} \times \mathcal{N})$; therefore the orthogonal projection v of F onto V exists. As an orthogonal projection, v is characterized by the relation

$$
\mathbb{E}\left[\left(F - \int_0^1 \dot{v}_0\, dW + \dot{v}_1\, d(N - t) \right) \left(\int_0^1 \dot{w}_0\, dW + \dot{w}_1\, d(N - t) \right) \right] = 0,
$$

for all $w \in \mathcal{P}$, and thus

$$
\mathbb{E}\left[F \int_0^1 \dot{w}_0\, dW + \dot{w}_1\, d(N - t) \right] = \mathbb{E}\left[\int_0^1 (\dot{v}_0\, \dot{w}_0 + \dot{v}_1\, \dot{w}_1)\, dt \right]
$$

$$
= \mathbb{E}\left[\int_0^1 \dot{w}_0(t)\, D_{t,0} F + \dot{w}_1(t)\, D_{t,1} F\, dt \right];
$$

for the last equality the formula of integration by parts (8.21) has been used. Taking the difference of the two members we get

$$
0 = \mathbb{E}\left[\int_0^1 \dot{w}_0(t)\, (D_{t,0} F - \dot{v}_0(t)) + \dot{w}_1(t)\, (D_{t,1} F - \dot{v}_1(t))\, dt \right]. \quad \square
$$

In general, mean-variance hedging is not exact hedging and therefore jump markets are not complete. The important question of completeness of some specific jump markets is discussed in [31, 64, 67, 69, 132, 161]. In some cases exact hedging may be realized by adding certain other martingales to the basic martingales W and $N(t) - t$.

A

Volatility Estimation by Fourier Expansion

Volatility is a key parameter in financial engineering. The classical Black–Scholes model assumes that volatilities of historical processes driven by the logarithm of the prices are constant. Empirical evidence has shown that this hypothesis is too restrictive. Time-varying volatilities are therefore a first step to adjust the Black–Scholes model to real data.

Estimation of volatility variations in the long range (month scale) is generally done by assuming an a priori model of stochastic volatility. The choice of this model is a difficult matter; its calibration is done by fitting the unknown parameters to data through some kind of Zakai filtering procedure; the evolution of historical volatilities is fitted by the solution of an SDE driven by unknown Brownian motions.

Experimental evidence of market evolution at an intra-day scale indicates that there exists no general model reasonably fitting data: as a consequence, one has to switch to *non-parametric Statistics*. In the usual sense statistics focuses on information extracted from the data of a population of different market evolutions. It should be emphasized that our method is designed to obtain results from the *observation of a single market evolution*. Such an approach is not out of reach for highly traded assets on which several thousand quotations can be made in a single day. The "statistical population" then has to be considered as the mass of information collected in time. As averaging principles are behind any statistical study, it is clear that volatility can only be approached on a time scale larger than the average frequency of the quotation data flow.

A possible procedure could be to split the time into subsequent intervals I_k, each containing at least a certain number of quotations. Then one may compute a volatility function σ^2 in the following way: σ^2 will be constant in each of the I_k and its value on I_k will be the mean quadratic variation of the observed quotations on I_k. This methodology proceeds by interval splitting; the drawback of information splitting procedures is a well-known fact in statistics.

The methodology of Fourier series proposed here can be considered as a refinement of this crude methodology, avoiding any information splitting and nevertheless being able to decipher abrupt variations in time of the historical volatility.

Another advantage of the Fourier series approach is the fact that the volatility is constructed as a *function*. It is therefore possible to iterate the procedure and to compute the cross-correlation between price and volatility; in a stable market this cross-correlation is negative; therefore an observed positive correlation monitors some market instability. A more advanced indicator of market stability is the *price-volatility feedback rate*, introduced in [19] and discussed in Chap. 3. A triple iteration of the Fourier series algorithm leads to a pathwise computation of this price-volatility feedback rate.

A common assumption in mathematical finance is that the time evolution of the price of an asset is a *semimartingale* of the form

$$dp = \sigma(t, W) \, dW + b(t, W) \, dt$$

where W is a Brownian motion. This assumption will be the only *a priori* assumption on which our non-parametric estimation will be based.

Using Itô calculus the volatility of the price process p is obtained by

$$\text{Vol}(p)(t) \equiv \sigma^2(t) = \lim_{\varepsilon \to 0+} \mathbb{E}^{\mathcal{N}_t} \left[\frac{(p(t + \varepsilon) - p(t))^2}{\varepsilon} \right].$$

However this formula cannot be used for numerical determination of the volatility; effectively only a single path of the market evolution is observed and the conditional expectation $\mathbb{E}^{\mathcal{N}_t}$ cannot be deduced from the observations.

As a consequence, estimation of the volatility is mainly obtained through the quadratic variation formula. In fact, there is a *pathwise formula*, essentially due to Norbert Wiener, which states that

$$\text{QV}\big|_{t_0}^{t_1}(p) = \int_{t_0}^{t_1} \sigma^2(s) \, ds$$

where the quadratic variation QV of p is given by

$$\text{QV}\big|_{t_0}^{t_1}(p) = \lim_{n \to \infty} \sum_{0 \le k < (t_1 - t_0)2^n} \left[p(t_0 + (k+1)2^{-n}) - p(t_0 + k2^{-n}) \right]^2.$$

Three bottlenecks appear when implementing this formula:

1. The price data are measured at *tick times* which are not equally spaced;
2. the computation of the volatility involves a numerical derivation;
3. the limit $n \to \infty$ cannot be effectively realized.

The Fourier series approach has the advantage of eliminating completely the second bottleneck and of smoothing the first one. The third bottleneck is not substantially changed; indeed the hypothesis that price evolution follows a semimartingale is only an approximation; the way to overcome this bottleneck is to approach limits as asymptotic series.

The computation of volatilities through harmonic analysis methods, first introduced by [145] has been used in [19–21]; see also [7, 60, 91–94, 105, 106].

In Sect. A.1 we shall prove an identity which gives an exact expression for the Fourier expansion of the volatility $\sigma^2(W, \cdot)$ in terms of the Fourier expansion of the price $p(\cdot)$. Section A.2 discusses the numerical implementation of this method.

A.1 Fourier Transform of the Volatility Functor

To a given function ϕ on the circle S^1 we associate its Fourier transform defined on the group of integers $\mathbb{Z}$ by the formula

$$\mathscr{F}(\phi)(k) = \frac{1}{2\pi} \int_0^{2\pi} \phi(\vartheta) \exp(-ik\vartheta) \, d\vartheta, \quad k \in \mathbb{Z} .$$

Given two functions Φ, Ψ on the integers, their Bohr convolution is defined as

$$(\Phi * \Psi)(k) = \lim_{N \to \infty} \frac{1}{2N+1} \sum_{s=-N}^{N} \Phi(s)\Psi(k-s) . \tag{A.1}$$

We denote by (H) the following hypothesis on the process p:

$$(\mathbf{H}) \qquad p(t) = p(0) + \int_0^t \sigma(s, W) \, dW(s) + \int_0^t b(s, W) \, ds$$

where σ is an adapted function, b is not necessarily adapted, both functions being bounded: $|b| + |\sigma| \le c$.

Theorem A.1. *Let p be a process satisfying assumption* (H) *along with the condition $p(0) = p(2\pi)$. Then the following formula holds:*

$$\frac{1}{2\pi} \mathscr{F}(\mathrm{Vol}(p)) = \Phi * \Psi, \quad \text{where } \Phi = \mathscr{F}(dp), \quad \bar{\Psi}(-k) = \mathscr{F}(dp)(k) ; \tag{A.2}$$

the equality holds in probability, which means that the limit in (A.1) *exists in probability.*

Remark A.2. When $p(0) \ne p(2\pi)$, we may introduce

$$\tilde{p}(t) = p(t) - \frac{p(2\pi) - p(0)}{2\pi} t .$$

Theorem A.1 then allows to compute the volatility of $\tilde{p}$. Finally we remark that p and $\tilde{p}$ have the same volatility.

Proof (of Theorem A.1). We give a sketch of the proof; details can be found in [145]. Since the drift b does not contribute to the quadratic variation, without loss of generality, we may suppose that $b = 0$; then p is a semimartingale.

We shall pursue the proof supposing that $\sigma(t, W)$ does not depend on W; the price to pay to treat this dependence are iterated Itô integrals realizing chaos expansion. Note that in the case where $\sigma(t)$ is independent of W, Wiener stochastic integrals are sufficient for the proof.

We introduce the complex martingales

$$\Gamma_k(t) := \frac{1}{2\pi} \int_0^t \sigma(s) \exp(-iks) \, dW(s), \quad k \in \mathbb{Z}.$$

By Itô calculus

$$\Gamma_k(2\pi)\Gamma_r(2\pi) = \frac{1}{2\pi}\mathscr{F}(\sigma^2)(k+r) + R(k,r)$$

where

$$R(k,r) := \int_0^{2\pi} (\Gamma_k \, d\Gamma_r + \Gamma_r \, d\Gamma_k).$$

Fix an integer $N \geq 1$ and define

$$\gamma_q(N) = \frac{1}{2N+1} \sum_{s=-N}^{N} \Gamma_{q+s}(2\pi)\Gamma_{-s}(2\pi).$$

By (A.1) we conclude that

$$\lim_{N \to \infty} \gamma_q(N) = (\Phi * \Psi)(q)$$

where $\Phi = \mathscr{F}(dp)$ and $\bar{\Psi}(-k) = \mathscr{F}(dp)(k)$. On the other hand, by the above product formula,

$$\gamma_q(N) = \frac{1}{2\pi}\mathscr{F}(\sigma^2)(q) + R_N,$$

where

$$R_N := \frac{1}{4\pi^2} \iint_{0 < t_1 < t_2 < 2\pi} D_N(t_1 - t_2) \left(e^{iqt_1} + e^{iqt_2}\right) \sigma(t_1)\sigma(t_2) \, dW(t_1) \, dW(t_2),$$

$$D_N(t) := \frac{1}{2N+1} \sum_{s=-N}^{N} e^{ist} = \frac{1}{2N+1} \frac{\sin(N + \frac{1}{2})t}{\sin \frac{t}{2}}.$$

We have

$$\int_0^{2\pi} |D_N|^2 \, dt = \frac{1}{2N+1}.$$

Using the energy identity for iterated stochastic integrals, we get

$$\mathbb{E}[R_N^2] \leq \frac{c}{2N+1}.$$

A disadvantage of the approximation procedure defined by (A.1) and (A.2) is that positivity of the volatility can be lost in the approximation. We shall modify slightly the procedure by taking advantage of the classical result:

Lemma A.3. *Given a function* $\Phi(k)$*, let* $\tilde{\Phi}(k) = \Phi(-k)$*. Then the convolution product* $\Phi * \tilde{\Phi}$ *has a positive Fourier transform:*

$$\mathscr{F}(\Phi * \tilde{\Phi}) = |\mathscr{F}(\Phi)|^2 .$$

We want to implement Lemma A.3 in real terms. Suppose that the Fourier coefficients $a_k(p)$, $b_k(p)$ are computed:

$$a_k(p) = \frac{1}{2\pi} \int_0^{2\pi} \cos(kt)p(t)dt, \quad b_k(p) = \frac{1}{2\pi} \int_0^{2\pi} \sin(kt)p(t) \, dt .$$

Then the Fourier coefficients of the derivative of p in the sense of distribution are given by

$$a_0(dp) = 0, \quad a_k(dp) = kb_k(p), \quad b_k(dp) = -ka_k(p), \quad k > 0 .$$

We define the prolongation to all integers k by parity for a_k and by imparity for b_k; more precisely let

$$a_0^* = b_0^* = 0, \quad a_k^* = \begin{cases} a_k(dp) & \text{for } k > 0 \\ a_{-k}(dp) & \text{for } k < 0 \end{cases} \quad \text{and} \quad b_k^* = \begin{cases} b_k(dp) & \text{for } k > 0 \\ -b_{-k}(dp) & \text{for } k < 0. \end{cases}$$

Theorem A.4. *Let* p *be a process satisfying hypothesis* **(H)***. For* $0 \leq q \leq 2N$ *where* N *is a positive integer, let*

$$\alpha_q(N) = \frac{1}{2N+1} \sum_{s=-N}^{N-q} (a_{q+s}^* a_s^* + b_{q+s}^* b_s^*) ,$$

$$\beta_q(N) = \frac{1}{2N+1} \sum_{s=-N}^{N-q} (-a_{q+s}^* b_s^* + b_{q+s}^* a_s^*) .$$

Then the trigonometric polynomial with $\alpha.(N), \beta.(N)$ *as coefficients is positive. Denote by* $a_q(\sigma^2)$*,* $b_q(\sigma^2)$ *the Fourier coefficients of* σ^2*. Then, for any fixed* $q \geq 0$*, as* $N \to \infty$*, the following convergence in probability holds true:*

$$\lim_N \alpha_q(N) = \frac{1}{\pi} a_q(\sigma^2), \quad \lim_N \beta_q(N) = \frac{1}{\pi} b_q(\sigma^2) . \tag{A.3}$$

Proof. We may again suppose that $b = 0$. We confine ourselves to the case where σ is a deterministic function of time; then p is a Gaussian process and the Fourier coefficients of its differentials are Gaussian variables. The covariances

$$\mathbb{E}\big[a_k(dp)a_s(dp)\big] = \frac{1}{2\pi}(a_{k+s}(\sigma^2) + a_{k-s}(\sigma^2))$$

$$\mathbb{E}\big[a_k(dp)b_s(dp)\big] = \frac{1}{2\pi}(b_{s+k}(\sigma^2) + b_{s-k}(\sigma^2))$$

$$\mathbb{E}\big[b_k(dp)b_s(dp)\big] = \frac{1}{2\pi}(-a_{k+s}(\sigma^2) + a_{k-s}(\sigma^2))$$

give

$$\mathbb{E}\big[a_{q+s}(dp)a_s(dp) + b_{q+s}(dp)b_s(dp)\big] = \frac{1}{\pi}a_q(\sigma^2)$$

$$\mathbb{E}\big[b_{q+s}(dp)a_s(dp) - b_s(dp)a_{q+s}(dp)\big] = \frac{1}{\pi}b_q(\sigma^2) \ .$$

Therefore

$$\pi\,\mathbb{E}[\alpha_q(N)] = \left(1 - \frac{q}{2N+1}\right)a_q(\sigma^2) \ ,$$

$$\pi\,\mathbb{E}[\beta_q(N)] = \left(1 - \frac{q}{2N+1}\right)b_q(\sigma^2),$$

and

$$\lim_N \mathbb{E}[\alpha_q(N)] = \frac{1}{\pi}a_q(\sigma^2), \quad \lim_N \mathbb{E}[\beta_q(N)] = \frac{1}{\pi}b_q(\sigma^2) \ .$$

Fix an integer $N \geq 1$ and define the random function

$$\Phi_N(k) = 1_{[-N,N]}(k)\,\Gamma_k, \quad \Gamma_k := a_k(dp) + ib_k(dp) \ .$$

Then, by Lemma A.3, the Fourier transform of the convolution $\Phi_N * \tilde{\Phi}_N$ is positive. Define for $q \in \{-2N, \ldots, 2N\}$,

$$\gamma_q(N) = \frac{1}{2N+1}\sum_s \Gamma_{q+s}\Gamma_{-s}\,1_{[-N,N]}(s)\,1_{[-N,N]}(q+s)$$

$$= \frac{1}{2N+1}(\Phi_N * \tilde{\Phi}_N)(q) \ .$$

Therefore $\gamma_q(N) = \alpha_q(N) + \beta_q(N)$, and then,

$$\mathbb{E}[|\gamma_q(N)|^2] - |\mathbb{E}[\gamma_q(N)]|^2 \leq \frac{3}{(2\pi)^4(2N+1)^2}\sum_{s_1,s_2:\ |s_i|\leq N}|c_{s_1-s_2}(\sigma)|^2$$

$$\leq \frac{1}{(2\pi)^3(2N+1)}\,\|\sigma\|_{L^2}^2 \to 0 \ .$$

A.2 Numerical Implementation of the Method

Construction of an Interpolating Trigonometric Polynomial

We rescale the day to $[0, 2\pi]$ by putting $T = (t_n - t_1)/2\pi$. Suppose that the price is known at a series of intermediate times t_k. Choosing $N \approx n$, the number of data items obtained during the day, we want to interpolate the series of tick prices by a price function $p(t)$ of the form

$$p(t) - p(t_1) - a\frac{t - t_1}{T} \simeq a_0 + \sum_{k=1}^{N}(a_k \cos k\vartheta + b_k \sin k\vartheta), \quad \vartheta = \frac{t - t_1}{T} , \quad (A.4)$$

where $a = p(t_n) - p(t_1)$ and where $\simeq$ means that we use a truncation of the Fourier series of the l.h.s. at step N. We take the function p to be constant in each interval and continuous from the right. We differentiate p once and find

$$p' = \sum_{j>1}(p(t_j) - p(t_{j-1}))\,\delta_{\vartheta_j}$$

where δ_x denotes the Dirac mass at point x. Then the Fourier coefficients of p are obtained via integration by parts on the interval $[\vartheta_{j-1}, \vartheta_j]$ where the function takes the constant value $p(t_{j-1})$:

$$\int_{\vartheta_{j-1}}^{\vartheta_j} p(\vartheta) \cos k\vartheta \, d\vartheta = \frac{1}{k} p(t_{j-1})(\sin k\vartheta_j - \sin k\vartheta_{j-1}) ,$$

$$a_k(p) = \frac{1}{\pi}\sum_{j>1}(p(t_{j-1}) - p(t_j)) \sin k\vartheta_j ,$$

$$b_k(p) = -\frac{1}{\pi}\sum_{j>1}(p(t_{j-1}) - p(t_j)) \cos k\vartheta_j .$$

Filtering High Modes of Volatility

We get high modes up to the order $2N$. A classical trick in Fourier Analysis is to filter progressively high modes. For this purpose we introduce the two following kernels, the first being the Fejer kernel, the second a modification of the Fejer kernel:

$$\varphi_1(\lambda) = \sup\{1 - |\lambda|, 0\}, \quad \varphi_2(\lambda) = \frac{\sin^2 \lambda}{\lambda^2} . \quad (A.5)$$

As an approximation of the volatility we take

$$A_i(t) = \alpha_0 + \sum_{q=1}^{2N} \varphi_i(\delta q)\,(\alpha_q \cos q\vartheta + \beta_q \sin q\vartheta), \quad i = 1, 2 , \quad (A.6)$$

where δ is a parameter adapted to the scale which is expected to give an appropriate resolution of the volatility. The advantage of the kernel φ_1 is that it has a larger smoothing effect whereas, for the same value of δ, the kernel φ_2 may exhibit sharper instantaneous variations. As far as the numerical value of

δ is concerned, for both kernels it is advisable to take $\delta \geq 1/(2N)$; in addition, numerical tests are necessary to optimize the choice of δ.

According to well-known diffraction phenomena of Fourier series near the limits of a time window, accurate results cannot be obtained immediately after opening or shortly before closure of the market.

B

Strong Monte-Carlo Approximation
of an Elliptic Market

In this appendix we discuss the problem of how to establish a Monte-Carlo simulation to a given strictly elliptic second order operator with real coefficients which leads to good numerical approximations of the fundamental solution to the corresponding heat operator. We are searching for a "good" scheme having the following two properties:

- The scheme does not need any simulation of iterated Itô integrals.
- The scheme has the strong property of approximation of order 1.

The Romberg iteration of the Euler scheme, based on the asymptotic expansion of Talay–Tubaro [201], is classically used to obtain weak approximations of arbitrary high order. Our approximation of strong order 1 could be tried in problems related to barrier options where the Talay–Tubaro methodology is not immediately applicable.

The exposition here follows closely the reference [59]; we shall not enter into the geometric ideas behind the computations which will be presented as straightforward tricks. In fact these geometric ideas are from [57].

Experts in Numerical Analysis may be sceptical about the existence of any "good" scheme. We have to emphasize that our new point of view will be to think in terms of *elliptic operators*, instead of the usual point of view to work on a *fixed* SDE. Many SDEs can be associated to an elliptic operator; each choice corresponds to a parametrization by the Wiener space of the Stroock–Varadhan solution of the martingale problem.

For applications to finance all these parametrizations are equivalent. However there exists a unique parametrization leading to the "good" scheme.

Change of parametrization means to replace the d-dimensional standard Brownian motion W on the Wiener space $\mathscr{W}$ by an *orthogonal transform* $\tilde{W}$ whose Itô differential takes the form

$$d\tilde{W}_k = \sum_{j=1}^{d} \left(\Omega_W(t)\right)_j^k dW_j(t) \tag{B.1}$$

where $t \mapsto \Omega_W(t)$ is an adapted process with values in the group $O(d)$ of orthogonal matrices. We denote by $O(\mathscr{W})$ the family of all such orthogonal transforms which is isomorphic to the path space $\mathbb{P}(\mathrm{O}(d))$ on $\mathrm{O}(d)$. Pointwise multiplication defines a group structure on $\mathbb{P}(\mathrm{O}(d)) \simeq \mathrm{O}(\mathscr{W})$.

Given on $\mathbb{R}^d$ the data of $d+1$ smooth vector fields $A_0, A_1, \ldots, A_d$, we consider the Itô SDE

$$d\xi_W(t) = A_0(\xi_W(t))\,dt + \sum_{k=1}^{d} A_k(\xi_W(t))\,dW_k(t), \quad \xi_W(0) = \xi_0\,, \qquad \text{(B.2)}$$

where we assume ellipticity, that is, for any $\xi \in \mathbb{R}^d$ the vectors $A_1(\xi), \ldots, A_d(\xi)$ constitute a basis of $\mathbb{R}^d$; the components of a vector field U in this basis are denoted $\langle U, A_k \rangle_\xi$ which gives the decomposition $U(\xi) = \sum_{k=1}^{d} \langle U, A_k \rangle_\xi A_k(\xi)$. By *change of parametrization* we mean the substitution of W by $\tilde{W}$ in (B.2); we then get an Itô process in $\tilde{W}$.

The group $\mathrm{O}(\mathscr{W})$ operates on the set of elliptic SDEs on $\mathbb{R}^d$ and the orbits of this action are classified by the corresponding elliptic operators $\mathscr{L}$.

B.1 Definition of the Scheme $\mathscr{S}$

Denote by $t_\varepsilon := \varepsilon \times$ integer part of t/ε; we define our scheme by

$$Z_{W_\varepsilon}(t) - Z_{W_\varepsilon}(t_\varepsilon) = A_0(Z_{W_\varepsilon}(t_\varepsilon))\,(t - t_\varepsilon) \qquad \text{(B.3)}$$

$$+ \sum_k A_k(Z_{W_\varepsilon}(t_\varepsilon))\,(W_k(t) - W_k(t_\varepsilon))$$

$$+ \frac{1}{2}\sum_{k,s}(\partial_{A_k} A_s)(Z_{W_\varepsilon}(t_\varepsilon))\left\{(W_k(t) - W_k(t_\varepsilon))(W_s(t) - W_s(t_\varepsilon)) - \varepsilon\eta_k^s\right\}$$

$$+ \frac{1}{2}\sum_{k,s,i} A_i(Z_{W_\varepsilon}(t_\varepsilon))\,\langle [A_s, A_i], A_k \rangle_{Z_{W_\varepsilon}(t_\varepsilon)}$$

$$\left\{(W_k(t) - W_k(t_\varepsilon))(W_s(t) - W_s(t_\varepsilon)) - \varepsilon\eta_k^s\right\} \qquad \text{(B.4)}$$

where W is standard Brownian motion on $\mathbb{R}^d$, and η_k^s the Kronecker symbol defined by $\eta_k^s = 1$ if $k = s$ and zero otherwise. Denote by $\mathbb{P}(\mathbb{R}^d)$ *the path space* on $\mathbb{R}^d$, that is the Banach space of continuous maps from $[0, T]$ into $\mathbb{R}^d$, endowed with the sup norm: $\|p_1 - p_2\|_\infty = \sup_{t \in [0,T]} |p_1(t) - p_2(t)|_{\mathbb{R}^d}$. Fixing $\xi_0 \in \mathbb{R}^d$, let $\mathbb{P}_{\xi_0}(\mathbb{R}^d)$ be the subspace of paths starting from ξ_0. Given Borel measures ρ_1, ρ_2 on $\mathbb{P}(\mathbb{R}^d)$, denote by $\mathscr{M}(\rho_1, \rho_2)$ the set of measurable maps $\Psi\colon \mathbb{P}(\mathbb{R}^d) \to \mathbb{P}(\mathbb{R}^d)$ such that $\Psi_* \rho_1 = \rho_2$; the *Monge transport norm* (see [147, 211]) is defined as

$$d_{\mathscr{M}}(\rho_1, \rho_2) := \left[\inf_{\Psi \in \mathscr{M}(\rho_1,\rho_2)} \int \|\Psi(p)\|_\infty^2\,\rho_1(dp)\right]^{1/2}\,.$$

Theorem B.1. *Assume ellipticity and assume the vector fields A_k along with their first three derivatives to be bounded. Fix $\xi_0 \in \mathbb{R}^d$ and let $\rho_{\mathscr{L}}$ be the measure on $P_{\xi_0}(\mathbb{R}^d)$ defined by the solution of the Stroock–Varadhan martingale problem [196] for the elliptic operator:*

$$\mathscr{L} = \frac{1}{2} \sum_{k,\alpha,\beta} A_k^\alpha A_k^\beta D_\alpha D_\beta + \sum_\alpha A_0^\alpha D_\alpha \text{ where } D_\alpha = \partial/\partial\xi^\alpha .$$

Further let $\rho_{\mathscr{S}}$ be the measure obtained by the scheme $\mathscr{S}$ with initial value $Z_{W_\varepsilon}(0) = \xi_0$. Then

$$\limsup_{\varepsilon \to 0} \frac{1}{\varepsilon} d_{\mathscr{M}}(\rho_{\mathscr{L}}, \rho_{\mathscr{S}}) =: c < \infty . \tag{B.5}$$

Remark The proof of Theorem B.1 will provide an explicit transport functional Ψ_0 which puts the statement in a constructive setting; the constant c is effective.

B.2 The Milstein Scheme

The Milstein scheme for SDE (B.2) (cf. for instance [151] formula (0.23) or [110] p. 345; see also [139]) is based on the following stochastic Taylor expansion of A_k along the diffusion trajectory: $A_k(\xi_W(t)) = A_k(\xi_W(t_\varepsilon)) + \sum_j (\partial_{A_j} A_k)(\xi_W(t_\varepsilon))(W_j(t) - W_j(t_\varepsilon)) + O(\varepsilon)$, which leads to

$$\xi_{W_\varepsilon}(t) - \xi_{W_\varepsilon}(t_\varepsilon) = \sum_k A_k\big(\xi_{W_\varepsilon}(t_\varepsilon)\big)\big(W_k(t) - W_k(t_\varepsilon)\big) + (t - t_\varepsilon) A_0\big(\xi_{W_\varepsilon}(t_\varepsilon)\big)$$

$$+ \sum_{i,k} (\partial_{A_i} A_k)\big(\xi_{W_\varepsilon}(t_\varepsilon)\big) \int_{t_\varepsilon}^t \big(W_i(s) - W_i(t_\varepsilon)\big)\, dW_k(s);$$

the computation of $\int_{t_\varepsilon}^t (W_i(s) - W_i(t_\varepsilon)) dW_k(s)$ gives the Milstein scheme

$$\xi_{W_\varepsilon}(t) - \xi_{W_\varepsilon}(t_\varepsilon) = \sum_k A_k\big(\xi_{W_\varepsilon}(t_\varepsilon)\big)\big(W_k(t) - W_k(t_\varepsilon)\big) + (t - t_\varepsilon) A_0\big(\xi_{W_\varepsilon}(t_\varepsilon)\big)$$

$$+ \frac{1}{2} \sum_{i,k} (\partial_{A_i} A_k)\big(\xi_{W_\varepsilon}(t_\varepsilon)\big)\Big(\big(W_i(t) - W_i(t_\varepsilon)\big)\big(W_k(t) - W_k(t_\varepsilon)\big) - \varepsilon\eta_k^i\Big) + R$$

where

$$R = \sum_{i<k} [A_i, A_k](\xi_{W_\varepsilon}(t_\varepsilon)) \int_{t_\varepsilon}^t \big(W_i(s) - W_i(t_\varepsilon)\big)\, dW_k(s)$$

$$- \big(W_k(s) - W_k(t_\varepsilon)\big)\, dW_i(s) .$$

It is well known that the Milstein scheme has the following strong approximation property:

$$\mathbb{E}\left[\sup_{t\in[0,1]}\|\xi_W(t) - \xi_{W_\varepsilon}(t)\|^2\right] = O(\varepsilon^2) . \tag{B.6}$$

The numerical difficulty related to the Milstein scheme is how to achieve a fast simulation of R. The purpose of this work is to show that this simulation can be avoided by a change of parametrization.

B.3 Horizontal Parametrization

Given d independent vector fields $A_1, \ldots, A_d$ on $\mathbb{R}^d$, we take the vectors $A_1(\xi), \ldots, A_d(\xi)$ as basis at the point ξ; the functions $\beta_{k,s}^i$, called *structural functions*, are defined (see [57]) by

$$\beta_{k,\ell}^i(\xi) = \langle [A_k, A_\ell], A_i\rangle_\xi; \quad [A_k, A_\ell](\xi) = \sum_i \beta_{k,\ell}^i(\xi)\, A_i(\xi) .$$

The structural functions are antisymmetric with respect to the two lower indices. Consider the *connection functions*, defined from the structural functions by

$$\Gamma_{k,s}^i = \frac{1}{2}\left(\beta_{k,s}^i - \beta_{s,i}^k + \beta_{i,k}^s\right) . \tag{B.7}$$

Let $\Gamma_k = (\Gamma_{k,s}^i)_{1\leq i,s\leq d}$ be the $d\times d$ matrix obtained by fixing the index k in the three indices functions $\Gamma_{k,s}^i$. Then, by means of the antisymmetry of $\beta_{k,s}^i$ in the two lower indices, Γ_k is an antisymmetric matrix:

$$2(\Gamma_{k,s}^i + \Gamma_{k,i}^s) = \beta_{k,s}^i - \beta_{s,i}^k + \beta_{i,k}^s + \beta_{k,i}^s - \beta_{i,s}^k + \beta_{s,k}^i = 0 .$$

The matrix Γ_k operates on the coordinate vectors of the basis $A_s(\xi)$ via $\Gamma_k(A_s) = \sum_i \Gamma_{k,s}^i A_i$. This gives $\Gamma_k(A_s) - \Gamma_s(A_k) = [A_k, A_s]$; the ith component of the l.h.s. is $\frac{1}{2}(\beta_{k,s}^i - \beta_{s,i}^k + \beta_{i,k}^s - \beta_{s,k}^i + \beta_{k,i}^s - \beta_{i,s}^k) = \beta_{k,s}^i$. Let $\mathbb{M} = \mathbb{R}^d \times \mathscr{E}_d$ where $\mathscr{E}_d$ is the vector space of $d\times d$ matrices. Define on $\mathbb{M}$ vector fields $\tilde{A}_k$, $k = 1, \ldots, d$, as follows:

$$\tilde{A}_k(\xi, e) = \left(\sum_\ell e_k^\ell A_\ell(\xi), \mathcal{N}_k(\xi, e)\right),$$

$$(\mathcal{N}_k)_r^s(\xi, e) = -\sum_{\ell,\ell'} e_k^\ell e_r^{\ell'} \Gamma_{\ell,\ell'}^s(\xi), \quad \xi \in \mathbb{R}^d,\ e \in \mathscr{E}_d. \tag{B.8}$$

Denoting for a vector Z on $\mathbb{M}$ by Z^H its projection on $\mathbb{R}^d$, we have:

Proposition B.2. *The vector fields $\tilde{A}_k$ satisfy the relation* $[\tilde{A}_k, \tilde{A}_s]^H = 0$.

Proof. The horizontal component $[\partial_{\tilde{A}_k}\tilde{A}_s]^H$ is given by

$$[\partial_{\tilde{A}_k}\tilde{A}_s]^H = \sum_i \tilde{A}_k^i\, \partial_i \tilde{A}_s^H + \sum_q \left(\sum_{\alpha,\beta}(\mathcal{N}_k)_\beta^\alpha\, \partial_{\binom{\alpha}{\beta}} e_s^q\right) A_q$$

$$= \sum_i \left(\sum_\ell e_k^\ell A_\ell^i\right)\left(\sum_{\ell'} e_s^{\ell'} \partial_i A_{\ell'}\right) - \sum_{\alpha,\beta}\left(\sum_{\ell,\ell'} e_k^\ell e_\beta^{\ell'}\, \Gamma_{\ell,\ell'}^\alpha\right)\left(\sum_q \eta_\alpha^q \eta_\beta^s A_q\right)$$

$$= \sum_i \left(\sum_\ell e_k^\ell A_\ell^i\right)\left(\sum_{\ell'} e_s^{\ell'} \partial_i A_{\ell'}\right) - \sum_{\alpha,\ell,\ell'} e_k^\ell\, e_s^{\ell'}\, \Gamma_{\ell,\ell'}^\alpha A_\alpha,$$

using the fact that $\partial_{\binom{\alpha}{\beta}} e_s^q = \eta_\alpha^q \eta_\beta^s$. We finally get

$$[\partial_{\tilde{A}_k}\tilde{A}_s]^H = \sum_{\ell,\ell'} e_k^\ell e_s^{\ell'} \big[\partial_{A_\ell} A_{\ell'} - \sum_\alpha \Gamma_{\ell,\ell'}^\alpha A_\alpha\big].$$

Therefore the horizontal component of the commutator is

$$[\tilde{A}_k, \tilde{A}_s]^H = \sum_{\ell,\ell'} e_k^\ell e_s^{\ell'} [A_\ell, A_{\ell'}] - \sum_{\alpha,\ell,\ell'} e_k^\ell e_s^{\ell'} \left(\Gamma_{\ell,\ell'}^\alpha - \Gamma_{\ell',\ell}^\alpha\right) A_\alpha$$

which vanishes since $\Gamma_\ell(A_{\ell'}) - \Gamma_{\ell'}(A_\ell) = [A_\ell, A_{\ell'}]$. $\square$

Denote by e^T the transpose of the matrix e and let

$$\tilde{A}_0(\xi, e) = \left(A_0(\xi) - \frac{1}{2}eJ\right), \quad J := \sum_{k=1}^d \mathcal{N}_k^T \mathcal{N}_k.$$

Consider the following Itô SDE on the vector space $\mathbb{M}$:

$$dm_W = \sum_k \tilde{A}_k(m_W)\, dW_k + \tilde{A}_0(m_W)\, dt, \quad m_W(0) = (\xi_0, \mathrm{Id}). \tag{B.9}$$

Proposition B.3. *Denote* $m_W(t) = (\tilde{\xi}_W(t), e_W(t))$, *then* $e_W(t)$ *is an orthogonal matrix for* $t \geq 0$, *and for any* $f \in C^2(\mathbb{R}^d)$,

$$f(\tilde{\xi}_W(t)) - \int_0^t (\mathscr{L}f)(\tilde{\xi}_W(s))\, ds$$

is a local martingale.

Proof. We compute the stochastic differential of $e^T e$:

$$d(e^T e)_\ell^{\ell'} = \sum_k d(e_k^\ell e_k^{\ell'}) = -\sum_{m,k,p}\left(\sum_u e_k^\ell e_m^p e_k^u\, \Gamma_{p,u}^{\ell'} + \sum_v e_k^{\ell'} e_m^p e_k^v\, \Gamma_{p,v}^\ell\right) dW_m$$

$$+ \sum_k \left(e_k^\ell\, (\tilde{A}_0)_k^{\ell'} + e_k^{\ell'}\, (\tilde{A}_0)_k^\ell + \sum_{m,p,q,p',q'} e_m^p e_k^{p'} \Gamma_{p,p'}^{\ell'} e_m^q e_k^{q'} \Gamma_{q,q'}^\ell\right) dt,$$

where the last term of the drift comes from the Itô contraction $\sum_k de_k^\ell *$ $de_k^{\ell'} = \sum_k (\mathcal{N}_k^T \mathcal{N}_k)_\ell^{\ell'} dt = J_\ell^{\ell'} dt$. The first two terms of the drift are computed by using the definition of $\tilde{A}_0$: $\sum_k [e_k^\ell (\tilde{A}_0)_k^{\ell'} + e_k^{\ell'} (\tilde{A}_0)_k^\ell] = -\frac{1}{2} [(e^T e J)_\ell^{\ell'} + (e^T e J)_{\ell'}^\ell]$. Write $e^T e = \mathrm{Id} + \sigma$, then the drift takes the form $-(\sigma J + J\sigma)/2$. We compute the coefficient of dW_m:

$$-\sum_{k,p} \left(\sum_u e_k^\ell e_m^p e_k^u \, \Gamma_{p,u}^{\ell'} + \sum_v e_k^{\ell'} e_m^p e_k^v \, \Gamma_{p,v}^\ell \right)$$

$$= -\sum_p e_m^p \left(\sum_u (e^T e)_\ell^u \, \Gamma_{p,u}^{\ell'} + \sum_v (e^T e)_{\ell'}^v \, \Gamma_{p,v}^\ell \right).$$

Using the antisymmetry $\Gamma_{p,\ell}^{\ell'} = -\Gamma_{p,\ell'}^\ell$ we obtain

$$d\sigma_{\ell'}^\ell = -\sum_m dW_m \sum_p e_m^p \left(\sum_u \sigma_\ell^u \, \Gamma_{p,u}^{\ell'} + \sum_v \sigma_{\ell'}^v \, \Gamma_{p,v}^\ell \right) - \frac{1}{2} (\sigma J + J\sigma)_{\ell'}^\ell \, dt .$$

(B.10)

Equation (B.10), together with (B.9), gives an SDE with local Lipschitz coefficients for the triple $(\tilde{\xi}, e, \sigma)$; by uniqueness of the solution, as $\sigma(0) = 0$, we deduce that $\sigma(t) = 0$ for all $t \geq 0$. $\square$

In terms of the new $\mathbb{R}^d$-valued Brownian motion $\tilde{W}$ defined by $d\tilde{W}_k(t) := \sum_\ell (e_W(t))_\ell^k dW_\ell$, we have

$$d\tilde{\xi}_W = \sum_k A_k(\tilde{\xi}_W(t)) \, d\tilde{W}_k(t) + A_0(\tilde{\xi}_W(t)) \, dt .$$

(B.11)

B.4 Reconstruction of the Scheme $\mathscr{S}$

We want to prove that our scheme $\mathscr{S}$ is *essentially* the projection of the Milstein scheme $(\tilde{\xi}_{W_\varepsilon}, e_{W_\varepsilon})$ for the solution $m_W = (\tilde{\xi}_W, e_W)$ of the SDE (B.9). In order to write the first component $\tilde{\xi}_{W_\varepsilon}$ we have to compute the horizontal part of $\partial_{\tilde{A}_k} \tilde{A}_j$, which has been done in the proof of Proposition B.2; we get

$$\tilde{\xi}_{W_\varepsilon}(t) - \tilde{\xi}_{W_\varepsilon}(t_\varepsilon) = A_0(\tilde{\xi}_{W_\varepsilon}(t_\varepsilon)) (t - t_\varepsilon) + \sum_{k,\ell} (e_{W_\varepsilon}(t_\varepsilon))_k^\ell A_\ell(\tilde{\xi}_{W_\varepsilon}(t_\varepsilon)) \, \Delta(W_k)$$

$$+ \frac{1}{2} \sum_{k,j} \left\{ \sum_{\ell,\ell'} (e_{W_\varepsilon}(t_\varepsilon))_k^\ell (e_{W_\varepsilon}(t_\varepsilon))_j^{\ell'} \left(\partial_{A_\ell} A_{\ell'} - \sum_i \Gamma_{\ell,\ell'}^i A_i \right) (\tilde{\xi}_{W_\varepsilon}(t_\varepsilon)) \right\}$$

$$\times \left(\Delta(W_k)\Delta(W_j) - \varepsilon \eta_k^j \right)$$

where $\Delta(W_k) = W_k(t) - W_k(t_\varepsilon)$. By (B.6)

$$\mathbb{E} \left[\sup_{t \in [0,1]} \|e_W(t) - e_{W_\varepsilon}(t)\|^2 \right] \leq c\varepsilon^2, \quad \mathbb{E} \left[\sup_{t \in [0,1]} \|\tilde{\xi}_W(t) - \tilde{\xi}_{W_\varepsilon}(t)\|^2 \right] \leq c\varepsilon^2 .$$

(B.12)

Consider the new process $\xi_W^\sharp$ defined by

$$\xi_W^\sharp(t) - \xi_W^\sharp(t_\varepsilon) = A_0\big(\xi_W^\sharp(t_\varepsilon)\big)\,(t - t_\varepsilon) + \sum_{k,\ell} \big(e_W(t_\varepsilon)\big)_k^\ell\, A_\ell\big(\xi_W^\sharp(t_\varepsilon)\big)\,\Delta(W_k)$$

$$+ \frac{1}{2}\sum_{k,j}\left\{\sum_{\ell,\ell'} \big(e_W(t_\varepsilon)\big)_k^\ell \big(e_W(t_\varepsilon)\big)_j^{\ell'} \left(\partial_{A_\ell} A_{\ell'} - \sum_i \Gamma_{\ell,\ell'}^i A_i\right)\big(\xi_{W_\varepsilon}^\sharp(t_\varepsilon)\big)\right\}$$

$$\times \big(\Delta(W_k)\Delta(W_j) - \varepsilon\eta_k^j\big)\,.$$

Lemma B.4. *The process $\xi_W^\sharp$ has the same law as the process Z_{W_ε} defined in* (B.4).

Proof. By Proposition B.3, $\hat{W}_\ell(t) - \hat{W}_\ell(t_\varepsilon) := \sum_k \big(e_W(t_\varepsilon)\big)_k^\ell (W_k(t) - W_k(t_\varepsilon))$ are the increments of an $\mathbb{R}^d$-valued Brownian motion $\hat{W}$; we get

$$\xi_W^\sharp(t) - \xi_W^\sharp(t_\varepsilon) = A_0\big(\xi_W^\sharp(t_\varepsilon)\big)\,(t - t_\varepsilon)$$

$$+ \sum_k A_k\big(\xi_W^\sharp(t_\varepsilon)\big)\big(\hat{W}_k(t) - \hat{W}_k(t_\varepsilon)\big)$$

$$+ \frac{1}{2}\sum_{k,s}\left(\partial_{A_k} A_s - \sum_i \Gamma_{k,s}^i A_i\right)\big(\xi_W^\sharp(t_\varepsilon)\big)$$

$$\times \Big(\big(\hat{W}_k(t) - \hat{W}_k(t_\varepsilon)\big)\big(\hat{W}_s(t) - \hat{W}_s(t_\varepsilon)\big) - \varepsilon\eta_k^s\Big)\,.$$

By (B.7), we have $2\sum_i \Gamma_{k,s}^i A_i = [A_k, A_s] + \sum_i\big(\langle[A_i, A_s], A_k\rangle A_i + \langle[A_i, A_k], A_s\rangle A_i\big)$, where the first term is antisymmetric in k, s and does not contribute; the remaining sum is symmetric in k, s. Thus we get

$$-\sum_{i,k,s} \Gamma_{k,s}^i A_i\,\Delta(\hat{W}_k)\Delta(\hat{W}_s) = \sum_{i,k,s} A_i\langle[A_s, A_i], A_k\rangle\,\Delta(\hat{W}_k)\Delta(\hat{W}_s)$$

which proves Lemma B.4. $\square$

Lemma B.5. *We have* $\mathbb{E}\left[\sup_{t\in[0,1]} \|\xi_W^\sharp(t) - \tilde{\xi}_{W^\varepsilon}(t)\|_{\mathbb{R}^d}^2\right] \le c\varepsilon^2$.

Proof. The following method of introducing a parameter λ and differentiating with respect to λ, is steadily used in [146].

For $\lambda \in [0, 1]$, let $e^\lambda := \lambda e_W + (1 - \lambda)e_{W_\varepsilon}$ and define the process ξ_W^λ by

$$\xi_W^\lambda(t) - \xi_W^\lambda(t_\varepsilon) = A_0(\xi_W^\lambda(t_\varepsilon))(t - t_\varepsilon) + \sum_{k,\ell}(e_W^\lambda(t_\varepsilon))_k^\ell\, A_\ell(\xi_W^\lambda(t_\varepsilon))\Delta(W_k)$$

$$+ \frac{1}{2}\sum_{k,j}\left\{\sum_{\ell,\ell'}(e_W^\lambda(t_\varepsilon))_k^\ell (e_W^\lambda(t_\varepsilon))_j^{\ell'} \left(\partial_{A_\ell} A_{\ell'} - \sum_i \Gamma_{\ell,\ell'}^i A_i\right)(\xi_{W_\varepsilon}^\lambda(t_\varepsilon))\right\}$$

$$\times \big(\Delta(W_k)\Delta(W_j) - (t - t_\varepsilon)\eta_k^j\big)\,.$$

Let $u_W^\lambda := \partial \xi / \partial \lambda$; then $\xi_W^\sharp(t) - \tilde{\xi}_{W^\varepsilon}(t) = \int_0^1 u_W^\lambda(t)\, d\lambda$. Denote by A_k', $(\partial_{A_\ell} A_{\ell'})'$, $(\Gamma_{\ell,\ell'}^i A_i)'$ the matrices obtained by differentiating A_k, $\partial_{A_\ell} A_{\ell'}$, $\Gamma_{\ell,\ell'}^i A_i$ with respect to ξ, and consider the delayed matrix SDE

$$
dJ_{t \leftarrow t_0} = \Bigg[A_0'(\xi_W^\lambda(t_\varepsilon))\, dt + \sum_{k,\ell} (e_W^\lambda(t_\varepsilon))_k^\ell \, A_\ell'(\xi_W^\lambda(t_\varepsilon))\, dW_k
$$

$$
+ \frac{1}{2} \sum_{k,j} \Bigg\{ \sum_{\ell,\ell'} (e_W^\lambda(t_\varepsilon))_k^\ell \, (e_W^\lambda(t_\varepsilon))_j^{\ell'} \left((\partial_{A_\ell} A_{\ell'})' - \sum_i (\Gamma_{\ell,\ell'}^i A_i)' \right) (\xi_{W_\varepsilon}^\lambda(t_\varepsilon)) \Bigg\}
$$

$$
\times \left(\Delta(W_k)\, dW_j(t) + \Delta(W_j)\, dW_k(t) - \eta_k^j\, dt \right) \Bigg] J_{t \leftarrow t_0}
$$

with initial condition $J_{t_0 \leftarrow t_0} = \mathrm{Id}$. Then

$$
u_W^\lambda(T) = J_{T \leftarrow t_0} \int_0^T J_{t \leftarrow t_0}^{-1} \Bigg[\sum_{k,\ell} \left(e_W(t_\varepsilon) - e_{W_\varepsilon}(t_\varepsilon) \right)_k^\ell A_\ell(\xi_W^\lambda(t_\varepsilon))\, dW_k(t)
$$

$$
+ \sum_{k,j} \Bigg\{ \sum_{\ell,\ell'} \left(e_W(t_\varepsilon) - e_{W_\varepsilon}(t_\varepsilon) \right)_k^\ell \left(e_W^\lambda(t_\varepsilon) \right)_j^{\ell'} \left(\partial_{A_\ell} A_{\ell'} - \sum_i \Gamma_{\ell,\ell'}^i A_i \right) (\xi_{W_\varepsilon}^\lambda(t_\varepsilon)) \Bigg\}
$$

$$
\times \left(\Delta(W_k)\, dW_j(t) + \Delta(W_j)\, dW_k(t) - \eta_k^j\, dt \right) \Bigg]
$$

which along with (B.12) proves Lemma B.5. $\square$

C

Numerical Implementation of the Price-Volatility Feedback Rate

In this Section we present some numerical results illustrating the implementation in real time of the concepts introduced in Chapter 3.

Starting from a single market observation $S_W(t)$ over a well-defined interval of time, we let $x_W(t) := \log S_W(t)$. According to Theorem 3.4 the price-volatility feedback rate $\lambda(t)$ associated to $x_W(t)$ can be calculated in terms of the volatility function $A(t) = \text{vol}(x_W(t))$ of $x_W(t)$ and the cross-volatility functions $B(t)$, $C(t)$, defined by

$$dx * dx = A\,dt, \quad dA * dx = B\,dt, \quad dB * dx = C\,dt \ .$$

The functions $A(t)$, $B(t)$, $C(t)$ are estimated using the Fourier series method of Appendix A.

To illustrate the method, the values of $A(t)$, $B(t)$, $C(t)$ and $\lambda(t)$ have been calculated on IBM data for two trading days in the year 1999, see Figure C.2 (we acknowledge Olsen & Associates for the provision of the data sets). Since taking logarithms of the stock prices, mainly affects the scales of $A(t)$, $B(t)$, $C(t)$, but lets the shapes of the curves more or less invariant, the numerical computation has been done in terms of the stock quotations (without switching to the logarithmic scale).

For the sake of comparison, a simulation study was performed for geometric Brownian motion $S_W(t) = \exp(W(t) - t/2)$. Since the drift term $t/2$ in $x_W(t) = \log S_W(t) = W(t) - t/2$ does not affect the calculation of the volatilities, it has been ignored in the simulation. The first window in Figure C.1 shows a trajectory of $x_W(t) \equiv W(t)$ simulated on a scale with a resolution comparable to the IBM market data, where in the average a new quotation is made every 4 or 5 seconds during a trading day of 6.5 hours. On the basis of this simulated trajectory the volatilities were computed, adopting the same algorithm as for the IBM data, see Figure C.1 for the resulting curves.

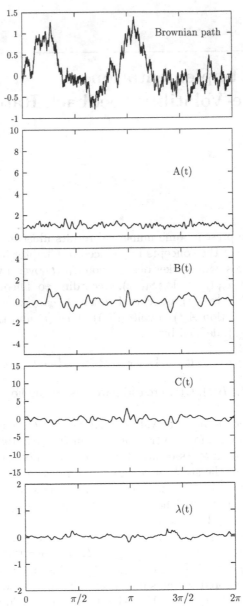

Fig. C.1. The price-volatility feedback rate for a simulated trajectory of (geometric) Brownian motion. As expected, the estimated volatility displays almost constant values close to 1, whereas the values of $B(t)$ and $C(t)$ oscillate near 0. As result, an almost vanishing price-volatility feedback rate is found: $\lambda(t) \approx 0$

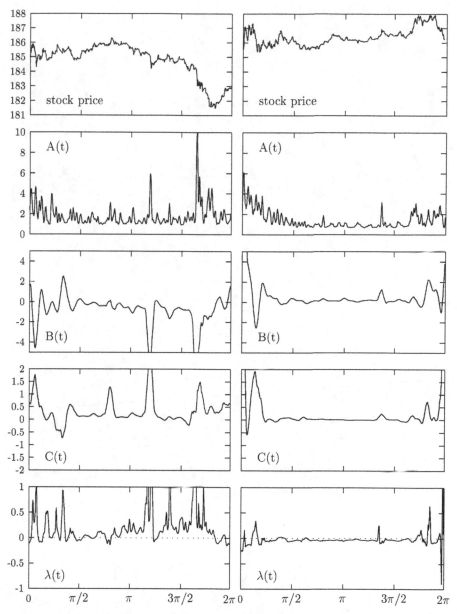

Fig. C.2. Daily values of $A(t), B(t), C(t), \lambda(t)$ on IBM data. The time windows $[0, 2\pi]$ correspond to two typical trading days in 1999 (6.5 hours each). Jan 4 (*left-hand side*) displays positivity of λ with large picks of A (instable market); April 9 (*right-hand side*) displays small and mainly negative values of λ along with a progressive damping of A (stable market). Graphics reproduced from [19] by courtesy of Blackwell Publishing Inc

References

1. Aase, K., Øksendal, B., Privault, N., Ubøe, J.: *White noise generalizations of the Clark–Haussmann–Ocone theorem with application to mathematical finance*, Finance Stoch. **4** (2000), no. 4, 465–496.
2. Aida, S., Kusuoka, S., Stroock, D.: *On the support of Wiener functionals*, Asymptotic problems in probability theory: Wiener functionals and asymptotics (Sanda/Kyoto, 1990), Pitman Res. Notes Math. Ser., vol. 284, Longman Sci. Tech., Harlow, 1993, pp. 3–34.
3. Airault, H.: *Perturbations singulières et solutions stochastiques de problèmes de D. Neumann-Spencer*, J. Math. Pures Appl. (9) **55** (1976), no. 3, 233–267.
4. Airault, H., Malliavin, P.: *Intégration géométrique sur l'espace de Wiener*, Bull. Sci. Math. (2) **112** (1988), no. 1, 3–52.
5. _____: *Backward regularity for some infinite dimensional hypoelliptic semigroups*, Stochastic analysis and related topics in Kyoto, Adv. Stud. Pure Math., vol. 41, Math. Soc. Japan, Tokyo, 2004, pp. 1–11.
6. Amendinger, J., Imkeller, P., Schweizer, M.: *Additional logarithmic utility of an insider*, Stochastic Process. Appl. **75** (1998), no. 2, 263–286.
7. Andersen, T., Bollerslev, T., Diebold, F.: *Parametric and Nonparametric Volatility Measurement*, Handbook of Financial Econometrics, North-Holland, 2004.
8. Arnaudon, M., Plank, H., Thalmaier, A.: *A Bismut type formula for the Hessian of a heat semigroup*, C. R. Acad. Sci. Paris, Ser. I, **336** (2003), 621–626.
9. Avellaneda, M., Gamba, R.: *Conquering the Greeks in Monte Carlo: efficient calculation of the market sensitivities and hedge-ratios of financial assets by direct numerical simulation*, Mathematical finance—Bachelier Congress, 2000 (Paris), Springer Finance, Springer, Berlin, 2002, pp. 93–109.
10. Azencott, R.: *Grandes déviations et applications*, Eighth Saint Flour Probability Summer School—1978 (Saint Flour, 1978), Lecture Notes in Math., vol. 774, Springer, Berlin, 1980, pp. 1–176.
11. Bally, V.: *On the connection between the Malliavin covariance matrix and Hörmander's condition*, J. Funct. Anal. **96** (1991), no. 2, 219–255.
12. Bachelier, L.: *Théorie de la spéculation*, Annales Scientifiques de l'Ecole Normale Supérieure 3e série, **17** (1900), 21–86.

13. Baldi, P., Caramellino, L., Iovino, M. G.: *Pricing general barrier options: a numerical approach using sharp large deviations*, Math. Finance **9** (1999), no. 4, 293–322.

14. Bally, V., L. Caramellino, L., Zanette, A.: *Pricing American options by Monte Carlo methods using a Malliavin Calculus approach*, Preprint INRIA, n. 4804, 2003.

15. Bally, V., Pagès, G., Printems, J.: *First-order schemes in the numerical quantization method*, Math. Finance **13** (2003), no. 1, 1–16.

16. Bally, V., Talay, D.: *The Euler scheme for stochastic differential equations: error analysis with Malliavin calculus*, Math. Comput. Simulation **38** (1995), no. 1-3, 35–41.

17. ———: *The law of the Euler scheme for stochastic differential equations. I. Convergence rate of the distribution function*, Probab. Theory Related Fields **104** (1996), no. 1, 43–60.

18. ———: *The law of the Euler scheme for stochastic differential equations. II. Convergence rate of the density*, Monte Carlo Methods Appl. **2** (1996), no. 2, 93–128.

19. Barrucci, E., Malliavin, P., Mancino, M. E., Renò, R., Thalmaier, A.: *The price-volatility feedback rate: an implementable mathematical indicator of market stability*, Math. Finance **13** (2003), no. 1, 17–35.

20. Barrucci, Renò, R.: *On measuring volatility of diffusion processes with high frequency data*, Economics Letters **74** (2002), 371–378.

21. ———: *On Measuring Volatility and the GARCH Forecasting Performance*, Journal of International Financial Markets, Institutions and Money **12** (2002), 183–200.

22. Bass R. F., Cranston, M.: *The Malliavin calculus for pure jump processes and applications to local time*, Ann. Probab. **14** (1986), no. 2, 490–532.

23. Baudoin, F.: *An introduction to the geometry of stochastic flows*, Imperial College Press, London, 2004.

24. Baudoin F., Teichmann J.: *Hypoellipticity in infinite dimensions and an application to interest rate theory*, Ann. Appl. Probab. **15** (2005), to appear.

25. Ben Arous, G.: *Développement asymptotique du noyau de la chaleur hypoelliptique hors du cut-locus*, Ann. Sci. École Norm. Sup. (4) **21** (1988), no. 3, 307–331.

26. ———: *Flots et séries de Taylor stochastiques*, Probab. Theory Related Fields **81** (1989), no. 1, 29–77.

27. Benhamou, E.: *An application of Malliavin calculus to continuous time asian options Greeks*, LSE Working Paper, 2000.

28. ———: *Smart Monte Carlo: various tricks using Malliavin calculus*, Quant. Finance **2** (2002), no. 5, 329–336.

29. ———: *Optimal Malliavin weighting function for the computation of the Greeks*, Math. Finance **13** (2003), no. 1, 37–53.

30. Benth, F. E., Dahl, L. O., Karlsen, K.H.: *On derivatives of claims in commodity and energy markets using a Malliavin approach*, Preprint, Dept. of Math., University of Oslo, 2002.

31. Benth, F. E., Di Nunno, G., Løkka, A., Øksendal, B., Proske, F.: *Explicit representation of the minimal variance portfolio in markets driven by Lévy processes*, Math. Finance **13** (2003), no. 1, 55–72.

32. Bermin, H.-P.: *A general approach to hedging options: applications to barrier and partial barrier options*, Math. Finance **12** (2002), no. 3, 199–218.
33. _____: *Hedging options: the Malliavin Calculus approach versus the Δ-hedging approach*, Math. Finance **13** (2003), no. 1, 73–84.
34. Bermin, H.-P., Kohatsu-Higa, A., Montero, M.: *Local Vega index and variance reduction methods*, Math. Finance **13** (2003), no. 1, 85–97.
35. Bernis, G., Gobet, E., Kohatsu-Higa, A.: *Monte Carlo evaluation of Greeks for multidimensional barrier and lookback options*, Math. Finance **13** (2003), no. 1, 99–113.
36. Bichteler, K., Gravereaux, J.-B., Jacod, J.: *Malliavin calculus for processes with jumps*, Stochastics Monographs, vol. 2, Gordon and Breach Science Publishers, New York, 1987.
37. Bismut, J.-M.: *Martingales, the Malliavin calculus and hypoellipticity under general Hörmander's conditions*, Z. Wahrsch. Verw. Gebiete **56** (1981), no. 4, 469–505.
38. Bismut, J.-M.: *Large deviations and the Malliavin calculus*, Progress in Mathematics, vol. 45, Birkhäuser Boston, Boston, MA, 1984.
39. Bogachev, V. I.: *Differentiable measures and the Malliavin calculus*, J. Math. Sci. (New York) **87** (1997), no. 4, 3577–3731.
40. _____: *Gaussian measures*, Mathematical Surveys and Monographs, vol. 62, American Mathematical Society, Providence, RI, 1998.
41. Bouchaud, J.-P., Cont, R., El-Karoui, N., Potters, M., Sagna, N.: *Phenomenology of the interest rate curve*, Appl. Math. Finance **6** (1999), no. 3, 209–232.
42. Bouchard, B., Ekeland, I., Touzi, N.: *On the Malliavin approach to Monte Carlo approximations for conditional expectations*, Finance Stoch. **8** (2004), no. 1, 45–71.
43. Bouleau, N.: *Error calculus for finance and physics: the language of Dirichlet forms*, de Gruyter Expositions in Mathematics, vol. 37, Walter de Gruyter & Co., Berlin, 2003.
44. Bouleau, N., Hirsch, F.: *Dirichlet forms and analysis on Wiener space*, de Gruyter Studies in Mathematics, vol. 14, Walter de Gruyter & Co., Berlin, 1991.
45. Brace, A., Musiela, M.: *A multifactor Gauss Markov implementation of Heath, Jarrow, and Morton*, Math. Finance **4** (1994), no. 3, 259–283.
46. Buckdahn R., Föllmer, H.: *A conditional approach to the anticipating Girsanov transformation*, Probab. Theory Related Fields **95** (1993), no. 3, 311–330.
47. Carlen, E. A., Pardoux, E.: *Differential calculus and integration by parts on Poisson space*, Stochastics, algebra and analysis in classical and quantum dynamics (Marseille, 1988), Math. Appl., vol. 59, Kluwer Acad. Publ., Dordrecht, 1990, pp. 63–73.
48. Carmona, R.: *Interest rate models: from parametric statistics to infinite dimensional stochastic analysis*, SIAM, to appear.
49. Carmona, R., Tehranchi, M.: *A characterization of hedging portfolios for interest rate contingent claims*, Ann. Appl. Probab. **14** (2004), no. 3, 1267–1294.
50. Castell, F.: *Asymptotic expansion of stochastic flows*, Probab. Theory Related Fields **96** (1993), no. 2, 225–239.
51. Clark, J. M. C.: *The representation of functionals of Brownian motion by stochastic integrals*, Ann. Math. Statist. **41** (1970), 1282–1295.
52. Cont, R.: *Modeling term structure dynamics: an infinite dimensional approach*, Int. J. Theor. Appl. Finance **8** (2005).

53. Cont, R., Tankov, P.: *Financial modelling with jump processes*, Chapman & Hall/CRC Financial Mathematics Series, Chapman & Hall/CRC, Boca Raton, FL, 2004.
54. Corcuera, J. M., Imkeller, P., Kohatsu-Higa, A., Nualart, D.: *Additional utility of insiders with imperfect dynamical information*, Finance Stoch. **8** (2004), no. 3, 437-450.
55. Corcuera, J.M., Nualart, D., Schoutens, W. *Completion of a Lévy market by power-jump assets*, Finance Stoch. **9** (2005), no. 1, 109–127.
56. Cruzeiro, A.-B.: *Équations différentielles sur l'espace de Wiener et formules de Cameron-Martin non-linéaires*, J. Funct. Anal. **54** (1983), no. 2, 206–227.
57. Cruzeiro, A.-B., Malliavin, P.: *Renormalized differential geometry on path space: structural equation, curvature*, J. Funct. Anal. **139** (1996), no. 1, 119–181.
58. _____: *Nonperturbative construction of invariant measure through confinement by curvature*, J. Math. Pures Appl. (9) **77** (1998), no. 6, 527–537.
59. Cruzeiro, A.-B., Malliavin, Thalmaier, A.: *Geometrization of Monte-Carlo numerical analysis of an elliptic operator: strong approximation*, C. R. Acad. Sci. Paris, Ser. I, 338 (2004) 481-486.
60. Cvitanić, J., Liptser, R., Rozovskii, B.: *A filtering approach to tracking volatility from prices observed at random times*, Preprint, 2004.
61. Cvitanić, J., Ma, J., Zhang, J.: *Efficient computation of Hedging portfolios for options with discontinuous payoffs*, Math. Finance **13** (2003), no. 1, 135–151.
62. Da Prato, G., Malliavin, P., Nualart, D.: *Compact families of Wiener functionals*, C. R. Acad. Sci. Paris Sér. I Math. **315** (1992), no. 12, 1287–1291.
63. Da Prato, G., Zabczyk, J.: *Stochastic equations in infinite dimensions*, Encyclopedia of Mathematics and its Applications, vol. 44, Cambridge University Press, Cambridge, 1992.
64. Davis, M., Johansson, M.: *Malliavin Monte Carlo Greeks for Jump Diffusions*, Preprint, Imperial College, 2004.
65. Delbaen, F., Schachermayer, W.: *A general version of the fundamental theorem of asset pricing*, Math. Ann. **300** (1994), no. 3, 463–520.
66. Dieudonné, J., Schwartz, L.: *La dualité dans les espaces $\mathcal{F}$ et ($\mathcal{LF}$)*, Ann. Inst. Fourier Grenoble **1** (1949), 61–101.
67. Di Nunno, G., Øksendal, B., Proske, F.: *White noise analysis for Lévy processes*, J. Funct. Anal. **206** (2004), no. 1, 109–148.
68. Doob, J. L.: *Boundary limit theorems for a half-space*, J. Math. Pures Appl. (9) **37** (1958), 385–392.
69. El-Khatib, Y., Privault, N.: *Hedging in complete markets driven by normal martingales*, Appl. Math. (Warsaw) **30** (2003), no. 2, 147–172.
70. _____: *Computations of Greeks in a market with jumps via the Malliavin calculus*, Finance Stoch. **8** (2004), no. 2, 161–179.
71. Fang, S., Malliavin, P.: *Stochastic analysis on the path space of a Riemannian manifold. I. Markovian stochastic calculus*, J. Funct. Anal. **118** (1993), no. 1, 249–274.
72. Filipović, D., Teichmann, J.: *Existence of invariant manifolds for stochastic equations in infinite dimension*, J. Funct. Anal. (2003), no. 2, 398–432.
73. _____: *Regularity of finite-dimensional realizations for evolution equations*, J. Funct. Anal. (2003), no. 2, 433–446.

74. Föllmer, H.: *Calcul d'Itô sans probabilités*, Seminar on Probability, XV (Univ. Strasbourg, Strasbourg, 1979/1980), Lecture Notes in Math., vol. 850, Springer, Berlin, 1981, pp. 143–150.
75. Föllmer, H., Imkeller, P.: *Anticipation cancelled by a Girsanov transformation: a paradox on Wiener space*, Ann. Inst. H. Poincaré Probab. Statist. **29** (1993), no. 4, 569–586.
76. Föllmer, H., Schied, A.: *Stochastic finance. An introduction in discrete time*, de Gruyter Studies in Mathematics, vol. 27, Walter de Gruyter & Co., Berlin, 2002.
77. Fouque, J.-P., Papanicolaou, G., Sircar, K. R.: *Derivatives in financial markets with stochastic volatility*, Cambridge University Press, Cambridge, 2000.
78. Fournié, E., Lasry, J. M., Lions, P.-L.: *Some nonlinear methods for studying far-from-the-money contingent claims*, Numerical methods in finance, Publ. Newton Inst., Cambridge Univ. Press, Cambridge, 1997, pp. 115–145.
79. Fournié, E., Lasry, J.-M., Lebuchoux, J., Lions, P.-L., Touzi, N.: *Applications of Malliavin calculus to Monte Carlo methods in finance*, Finance Stoch. **3** (1999), no. 4, 391–412.
80. Fournié, E., Lasry, J.-M., Lebuchoux, J., Lions, P.-L.: *Applications of Malliavin calculus to Monte-Carlo methods in finance. II*, Finance Stoch. **5** (2001), no. 2, 201–236.
81. Fournié, E., Lasry, J.-M., Touzi, N.: *Monte Carlo methods for stochastic volatility models*, Numerical methods in finance, Publ. Newton Inst., Cambridge Univ. Press, Cambridge, 1997, pp. 146–164.
82. Gaveau, B., Trauber, P.: *L'intégrale stochastique comme opérateur de divergence dans l'espace fonctionnel*, J. Funct. Anal. **46** (1982), no. 2, 230–238.
83. Gobet, E.: *Weak approximation of killed diffusion using Euler schemes*, Stochastic Process. Appl. **87** (2000), no. 2, 167–197.
84. _____: *Local asymptotic mixed normality property for elliptic diffusion: a Malliavin calculus approach*, Bernoulli **7** (2001), no. 6, 899–912.
85. _____: *LAN property for ergodic diffusions with discrete observations*, Ann. Inst. H. Poincaré Probab. Statist. **38** (2002), no. 5, 711–737.
86. _____: *Revisiting the Greeks for European and American options*, Stochastic processes and applications to mathematical finance. Proceedings Ritsumeikan Intern. Symposium, J. Akahori, S. Ogawa, S. Watanabe, Eds., World Scientific, 2004, pp. 53–71.
87. Gobet, E., Kohatsu-Higa, A.: *Computation of Greeks for barrier and look-back options using Malliavin calculus*, Electron. Comm. Probab. **8** (2003), 51–62.
88. Gobet, E., Munos, R.: *Sensitivity analysis using Itô–Malliavin calculus and martingales. Application to stochastic optimal control*, SIAM Journal on Control and Optimization (2005).
89. Grorud, A., Pontier, M.: *Comment détecter le délit d'initiés?* C. R. Acad. Sci. Paris Sér. I Math. **324** (1997), no. 10, 1137–1142.
90. _____: *Insider trading in a continuous time market model*, Int. J. Theor. Appl. Finance **1** (1998), no. 3, 331–347.
91. Hansen, P.R., Lunde, A.: *Consistent preordering with an estimated criterion function, with an application to the evaluation and comparison of volatility models*, Working paper 2003-01, Dept. of Economics, Brown University.
92. Hansen, P.R., Lunde, A., Nason, J.M.: *Choosing the best volatility models: the model confidence set approach*, Working paper 2003-05, Dept. of Economics, Brown University.

93. Hayashi, T., Kusuoka, S.: *Nonsynchronous covariation measurement for continuous semimartingales*, Preprint UTMS 2004-21, University of Tokyo.
94. Hayashi, T., Yoshida, N.: *On covariance estimation of nonsynchronously observed diffusion processes*, Bernoulli **11** (2005).
95. Heath, D., Jarrow, R., Morton, A.: *Bond pricing and the term structure of interest rates: A new methodology for contingent claims valuation*, Econometrica **60** (1992), no. 1, 77–105.
96. Ikeda, N., Watanabe, S.: *Stochastic differential equations and diffusion processes*, 2nd edn, North-Holland Mathematical Library, vol. 24, North-Holland Publishing Co., Amsterdam, 1989.
97. Imkeller, P.: *Enlargement of the Wiener filtration by an absolutely continuous random variable via Malliavin's calculus*, Probab. Theory Related Fields **106** (1996), no. 1, 105–135.
98. _____: *Enlargement of the Wiener filtration by a manifold valued random element via Malliavin's calculus*, Statistics and control of stochastic processes (Moscow, 1995/1996), World Sci. Publ., River Edge, NJ, 1997, pp. 157–171.
99. _____: *Malliavin's Calculus in insider models: additional utility and free lunches*, Math. Finance **13** (2003), no. 1, 153–169.
100. Imkeller, P., Pontier, M., Weisz, F.: *Free lunch and arbitrage possibilities in a financial market model with an insider*, Stochastic Process. Appl. **92** (2001), no. 1, 103–130.
101. Itô, K.: *A measure-theoretic approach to Malliavin calculus*, New trends in stochastic analysis (Charingworth, 1994), World Sci. Publishing, River Edge, NJ, 1997, pp. 220–287.
102. Itô, K., McKean, H. P. Jr.: *Diffusion processes and their sample paths*, Springer-Verlag, Berlin, 1974, Second printing, corrected, Die Grundlehren der mathematischen Wissenschaften, Band 125.
103. Jacod, J.: *Grossissement initial, hypothèse (H') et théorème de Girsanov*, Grossissements de filtrations: exemples et applications, Lecture Notes in Math., vol. 1118, Springer-Verlag, Berlin, 1985, pp. 15–35.
104. Janicki, A., Krajna, Ł.: *Malliavin calculus in construction of hedging portfolio for the Heston model of a financial market*, Demonstratio Math. **34** (2001), no. 2, 483–495.
105. Jungbacker, B., Koopman, S.J.: *Model-based measurement of actual volatility in high frequency data*, Preprint 2004, Dept. Econometrics, Free University Amsterdam.
106. Kanatami, T.: *Integrated volatility measuring from unevenly sampled observations*, Economics Bulletin **3** (2004), no. 36, 1–8.
107. Karatzas, I., Kou, S. G.: *Hedging American contingent claims with constrained portfolios*, Finance Stoch. **2** (1998), no. 3, 215–258.
108. Karatzas, I., Ocone, D. L., Li, J.: *An extension of Clark's formula*, Stochastics Stochastics Rep. **37** (1991), no. 3, 127–131.
109. Karatzas, I., Shreve, S. E.: *Methods of mathematical finance*, Applications of Mathematics, vol. 39, Springer-Verlag, New York, 1998.
110. Kloeden, P. E., Platen, E.: *Numerical solution of stochastic differential equations*, Applications of Mathematics (New York), vol. 23, Springer-Verlag, Berlin, 1992.
111. Kohatsu-Higa, A.: *Weak approximations. A Malliavin calculus approach*, Math. Comp. **70** (2001), no. 233, 135–172.

112. Kohatsu-Higa, A., Montero, M.: *Malliavin calculus applied to finance*, Phys. A **320** (2003), no. 1-4, 548–570.
113. _____: *Malliavin calculus in finance*, Handbook of computational and numerical methods in finance, Birkhäuser Boston, Boston, MA, 2004, pp. 111–174.
114. Kohatsu-Higa, A., Ogawa, S.: *Weak rate of convergence for an Euler scheme of nonlinear SDE's*, Monte Carlo Methods Appl. **3** (1997), no. 4, 327–345.
115. Kohatsu-Higa, A., Pettersson, R.: *Variance reduction methods for simulation of densities on Wiener space*, SIAM J. Numer. Anal. **40** (2002), no. 2, 431–450.
116. Kunita, H.: *Stochastic flows and stochastic differential equations*, Cambridge Studies in Advanced Mathematics, vol. 24, Cambridge University Press, Cambridge, 1990.
117. Kunitomo, N.,Takahashi, A.: *The asymptotic expansion approach to the valuation of interest rate contingent claims*, Math. Finance **11** (2001), no. 1, 117–151.
118. _____, *On validity of the asymptotic expansion approach in contingent claim analysis*, Ann. Appl. Probab. **13** (2003), no. 3, 914–952.
119. _____, *Applications of the asymptotic expansion approach based on Malliavin-Watanabe Calculus in financial problems*, Stochastic processes and applications to mathematical finance. Proceedings Ritsumeikan Intern. Symposium, J. Akahori, S. Ogawa, S. Watanabe, Eds., World Scientific, 2004, pp. 195–232.
120. Kusuoka, S.: *Approximation of expectation of diffusion process and mathematical finance*, Taniguchi Conference on Mathematics Nara '98, Adv. Stud. Pure Math., vol. 31, Math. Soc. Japan, Tokyo, 2001, pp. 147–165.
121. _____, *Malliavin calculus revisited*, J. Math. Sci. Univ. Tokyo **10** (2003), no. 2, 261–277.
122. _____, *Approximation of expectation of diffusion processes based on Lie algebra and Malliavin calculus*, Advances in mathematical economics. Vol. 6, Springer, Tokyo, 2004, pp. 69–83.
123. Kusuoka, S., Ninomiya, S.: *A new simulation method of diffusion processes applied to Finance*, Stochastic processes and applications to mathematical finance. Proceedings Ritsumeikan Intern. Symposium, J. Akahori, S. Ogawa, S. Watanabe, Eds., World Scientific, 2004, pp. 233–253.
124. Kusuoka, S., Stroock, D.: *Applications of the Malliavin calculus. II*, J. Fac. Sci. Univ. Tokyo Sect. IA Math. **32** (1985), no. 1, 1–76.
125. _____: *Applications of the Malliavin calculus. III*, J. Fac. Sci. Univ. Tokyo Sect. IA Math. **34** (1987), no. 2, 391–442.
126. _____: *Precise asymptotics of certain Wiener functionals*, J. Funct. Anal. **99** (1991), no. 1, 1–74.
127. _____: *Asymptotics of certain Wiener functionals with degenerate extrema*, Comm. Pure Appl. Math. **47** (1994), no. 4, 477–501.
128. Kusuoka, S., Yoshida, N.: *Malliavin calculus, geometric mixing, and expansion of diffusion functionals*, Probab. Theory Related Fields **116** (2000), no. 4, 457–484.
129. Lamberton, D., Lapeyre, B.: *Introduction to stochastic calculus applied to finance*, Chapman & Hall, London, 1996.
130. Ledoux, M.: *L'algèbre de Lie des gradients itérés d'un générateur markovien*, C. R. Acad. Sci. Paris Sér. I Math. **317** (1993), no. 11, 1049–1052.
131. León, J. A., Navarro, R., Nualart, D.: *An anticipating calculus approach to the utility maximization of an insider*, Math. Finance **13** (2003), no. 1, 171–185.
132. León, J. A., Solé, J. L., Utzet, F., Vives, J: *On Lévy processes, Malliavin calculus and market models with jumps*, Finance Stoch. **6** (2002), no. 2, 197–225.

133. Lions, P.-L.: *Optimal control of diffusion processes and Hamilton–Jacobi–Bellman equations. I. The dynamic programming principle and applications*, Comm. Partial Differential Equations **8** (1983), no. 10, 1101–1174.

134. _____: *Optimal control of diffusion processes and Hamilton–Jacobi–Bellman equations. II. Viscosity solutions and uniqueness*, Comm. Partial Differential Equations **8** (1983), no. 11, 1229–1276.

135. _____: *Optimal control of diffusion processes and Hamilton–Jacobi–Bellman equations. III. Regularity of the optimal cost function*, Nonlinear partial differential equations and their applications. Collège de France seminar, Vol. V (Paris, 1981/1982), Res. Notes in Math., vol. 93, Pitman, Boston, MA, 1983, pp. 95–205.

136. _____: *On mathematical finance*, Boll. Unione Mat. Ital. Sez. B Artic. Ric. Mat. (8) **3** (2000), no. 3, 553–572.

137. Lütkebohmert, E.: *An asymptotic expansion for a Black–Scholes type model*, Bull. Sci. Math. **128** (2004), no. 8, 661–685.

138. Lyons, T., Victoir, N.: *Cubature on Wiener space*, Proc. R. Soc. Lond. Ser. A Math. Phys. Eng. Sci. **460** (2004), no. 2041, 169–198.

139. Malliavin, P.: *Paramétrix trajectorielle pour un opérateur hypoelliptique et repère mobile stochastique*, C. R. Acad. Sci. Paris Sér. A-B 281 (1975), A241–A244.

140. _____: C^k-*hypoellipticity with degeneracy*, Stochastic analysis (Proc. Internat. Conf., Northwestern Univ., Evanston, Ill., 1978), Academic Press, New York, 1978, pp. 199–214, 327–340.

141. _____: *Stochastic calculus of variation and hypoelliptic operators*, Proceedings of the International Symposium on Stochastic Differential Equations (Res. Inst. Math. Sci., Kyoto Univ., Kyoto, 1976) (New York), Wiley, 1978, pp. 195–263.

142. _____: *Géométrie différentielle stochastique*, Séminaire de Mathématiques Supérieures, vol. 64, Presses de l'Université de Montréal, Montreal, 1978.

143. _____: *Integration and probability*, Graduate Texts in Mathematics, vol. 157, Springer-Verlag, New York, 1995.

144. _____: *Stochastic analysis*, Grundlehren der Mathematischen Wissenschaften, vol. 313, Springer-Verlag, Berlin, 1997.

145. Malliavin, P., Mancino, M. E.: *Fourier series method for measurement of multivariate volatilities*, Finance Stoch. **6** (2002), no. 1, 49–61.

146. Malliavin, P., Thalmaier, A.: *Numerical error for SDE: asymptotic expansion and hyperdistributions*, C. R. Math. Acad. Sci. Paris 336 (2003), 851–856.

147. Malrieu, F.: *Convergence to equilibrium for granular media equations and their Euler schemes*, Ann. Appl. Probab. 13 (2003), 540–560.

148. Mancino, M. E., Ogawa, S.: *Nonlinear feedback effects by hedging strategies*, Stochastic processes and applications to mathematical finance. Proceedings Ritsumeikan Intern. Symposium, J. Akahori, S. Ogawa, S. Watanabe, Eds., World Scientific, 2004.

149. McKean, H. P. Jr.: *Stochastic integrals*, Probability and Mathematical Statistics, No. 5, Academic Press, New York, 1969.

150. Meyer, P.-A.: *Transformations de Riesz pour les lois gaussiennes*, Seminar on probability, XVIII, Lecture Notes in Math., vol. 1059, Springer, Berlin, 1984, pp. 179–193.

151. Milstein, G. N.: *Numerical integration of stochastic differential equations*, Mathematics and its Applications, vol. 313, Kluwer Academic Publishers, Dordrecht, 1995, Translated and revised from the 1988 Russian original.

152. Musiela, M., Rutkowski, M.: *Martingale methods in financial modelling*, Applications of Mathematics, vol. 36, Springer-Verlag, Berlin, 1997.

153. Ninomiya, S.: *A partial sampling method applied to the Kusuoka approximation*, Monte Carlo Methods Appl. **9** (2003), no. 1, 27–38.

154. Ninomiya, S., Tezuka, S.: *Toward real-time pricing of complex financial derivatives*, Appl. Math. Finance **3** (1996), no. 1, 1–20.

155. Norris, J. R.: *Integration by parts for jump processes*, Séminaire de Probabilités, XXII, Lecture Notes in Math., vol. 1321, Springer, Berlin, 1988, pp. 271–315.

156. _____ : *Simplified Malliavin calculus*, Séminaire de Probabilités, XX, 1984/85, Lecture Notes in Math., vol. 1204, Springer, Berlin, 1986, pp. 101–130.

157. Nualart, D.: *Noncausal stochastic integrals and calculus*, Stochastic analysis and related topics (Silivri, 1986), Lecture Notes in Math., vol. 1316, Springer, Berlin, 1988, pp. 80–129.

158. _____ : *Une remarque sur le développement en chaos d'une diffusion*, Séminaire de Probabilités, XXIII, Lecture Notes in Math., vol. 1372, Springer, Berlin, 1989, pp. 165–168.

159. _____ : *The Malliavin calculus and related topics*, Probability and its Applications, Springer-Verlag, New York, 1995.

160. Nualart, D., Pardoux, É.: *Stochastic calculus with anticipating integrands*, Probab. Theory Related Fields **78** (1988), no. 4, 535–581.

161. Nualart, D., Schoutens, W.: *Chaotic and predictable representations for Lévy processes*, Stochastic Process. Appl. **90** (2000), no. 1, 109–122.

162. _____ , *Backward stochastic differential equations and Feynman–Kac formula for Lévy processes, with applications in finance*, Bernoulli **7** (2001), no. 5, 761–776.

163. Nualart D., Üstünel, A. S.: *Geometric analysis of conditional independence on Wiener space*, Probab. Theory Related Fields **89** (1991), no. 4, 407–422.

164. Nualart, D., Vives, J.: *Continuité absolue de la loi du maximum d'un processus continu*, C. R. Acad. Sci. Paris Sér. I Math. **307** (1988), no. 7, 349–354.

165. Nualart, D., Zakai, M.: *Generalized stochastic integrals and the Malliavin calculus*, Probab. Theory Relat. Fields **73** (1986), no. 2, 255–280.

166. _____ : *A summary of some identities of the Malliavin calculus*, Stochastic partial differential equations and applications, II (Trento, 1988), Lecture Notes in Math., vol. 1390, Springer, Berlin, 1989, pp. 192–196.

167. _____ : *On the relation between the Stratonovich and Ogawa integrals*, Ann. Probab. **17** (1989), no. 4, 1536–1540.

168. Ocone, D.: *Malliavin's calculus and stochastic integral representations of functionals of diffusion processes*, Stochastics **12** (1984), no. 3-4, 161–185.

169. _____ : *A guide to the stochastic calculus of variations*, Stochastic analysis and related topics (Silivri, 1986), Lecture Notes in Math., vol. 1316, Springer, Berlin, 1988, pp. 1–79.

170. Ocone, D., Karatzas, I.: *A generalized Clark representation formula, with application to optimal portfolios*, Stochastics Stochastics Rep. **34** (1991), no. 3-4, 187–220.

171. Ogawa, S.: *The stochastic integral of noncausal type as an extension of the symmetric integrals*, Japan J. Appl. Math. **2** (1985), no. 1, 229–240.

172. Øksendal, B.: *An Introduction to Malliavin Calculus with Applications to Economics*, Lecture Notes from a course given 1996 at the Norwegian School of Economics and Business Administration (NHH), 1997, NHH Preprint Series.

173. _____: *Stochastic differential equations. An introduction with applications*, 5th edn, Universitext, Springer-Verlag, Berlin, 1998.

174. Øksendal, B., Sulem, A.: *Applied stochastic control of jump diffusions*, Universitext, Springer-Verlag, Berlin, 2005.

175. Parthasarathy, K. R.: *Probability measures on metric spaces*, Probability and Mathematical Statistics, No. 3, Academic Press, New York, 1967.

176. Picard, J.: *Approximation of stochastic differential equations and application of the stochastic calculus of variations to the rate of convergence*, Stochastic analysis and related topics (Silivri, 1986), Lecture Notes in Math., vol. 1316, Springer, Berlin, 1988, pp. 267–287.

177. Pedersen, J.: *Convergence of strategies: an approach using Clark-Haussmann's formula*, Finance Stoch. **3** (1999), no. 3, 323–344.

178. Pisier, G.: *Riesz transforms: a simpler analytic proof of P.-A. Meyer's inequality*, Séminaire de Probabilités, XXII, Lecture Notes in Math., vol. 1321, Springer, Berlin, 1988, pp. 485–501.

179. Privault, N.: *Chaotic and variational calculus in discrete and continuous time for the Poisson process*, Stochastics Stochastics Rep. **51** (1994), no. 1-2, 83–109.

180. Privault, N., Wei, Xiao: *A Malliavin calculus approach to sensitivity analysis in insurance*, Insurance Math. Econom. **35** (2004), no. 3, 679–690.

181. Régnier, H.: *Calcul du prix et des sensibilités d'une option américaine par une méthode de Monte-Carlo*, Preprint, 2000.

182. _____: *Monte Carlo computations of American options via Malliavin calculus*, Preprint, 2003.

183. Russo, F, Vallois, P.: *Forward, backward and symmetric stochastic integration*, Probab. Theory Related Fields **97** (1993), no. 3, 403–421.

184. Schachermayer, W.: *Introduction to the mathematics of financial markets*, Lectures on probability theory and statistics (Saint-Flour, 2000), Lecture Notes in Math., vol. 1816, Springer, Berlin, 2003, pp. 107–179.

185. Serrat, A.: *A dynamic equilibrium model of international portfolio holdings*, Econometrica **69** (2001), no. 6, 1467–1489.

186. Shigekawa, I.: *Derivatives of Wiener functionals and absolute continuity of induced measures*, J. Math. Kyoto Univ. **20** (1980), no. 2, 263–289.

187. _____: *de Rham–Hodge–Kodaira's decomposition on an abstract Wiener space*, J. Math. Kyoto Univ. **26** (1986), no. 2, 191–202.

188. _____: *Stochastic analysis*, Translations of Mathematical Monographs, vol. 224, American Mathematical Society, Providence, RI, 2004, Translated from the 1998 Japanese original, Iwanami Series in Modern Mathematics.

189. Sircar, K. R., Papanicolaou, G. C.: *Stochastic volatility, smile & asymptotics*, Appl. Math. Finance **6**, no. 2, 107–145.

190. Stroock, D. W.: *The Malliavin calculus, a functional analytic approach*, J. Funct. Anal. **44** (1981), no. 2, 212–257.

191. _____: *The Malliavin calculus and its application to second order parabolic differential equations. I*, Math. Systems Theory **14** (1981), no. 1, 25–65.

192. _____: *The Malliavin calculus and its application to second order parabolic differential equations. II*, Math. Systems Theory **14** (1981), no. 2, 141–171.

193. _____: *The Malliavin calculus and its applications*, Stochastic integrals (Proc. Sympos., Univ. Durham, Durham, 1980), Lecture Notes in Math., vol. 851, Springer, Berlin, 1981, pp. 394–432.

194. _____: *Homogeneous chaos revisited*, Séminaire de Probabilités, XXI, Lecture Notes in Math., vol. 1247, Springer, Berlin, 1987, pp. 1–7.

195. _____: *An introduction to the analysis of paths on a Riemannian manifold*, Mathematical Surveys and Monographs, vol. 74, American Mathematical Society, Providence, RI, 2000.

196. Stroock, D.W., Varadhan, S.R.S.: *Multidimensional diffusion processes*, Grundlehren der Math. Wissenschaften, vol. 233, Springer-Verlag, Berlin, 1979.

197. Sugita, H.: *Sobolev spaces of Wiener functionals and Malliavin's calculus*, J. Math. Kyoto Univ. **25** (1985), no. 1, 31–48.

198. Takahashi, A.: *An Asymptotic Expansion Approach to Pricing Financial Contingent Claims*, Asia-Pacific Financial Markets **6** (1999), 115–151.

199. Takahashi, A., Yoshida, N.: *An Asymptotic Expansion Scheme for Optimal Investment Problems*, Statistical Inference for Stochastic Processes **7** (2004), 153–188.

200. _____: *Monte Carlo simulation with asymptotic method*, Journal of Japan Statistical Society, to appear.

201. Talay D., Tubaro, L.: *Expansion of the global error for numerical schemes solving stochastic differential equations*, Stochastic Anal. Appl. **8** (1990), no. 4, 483–509.

202. Talay, D., Zheng, Z.: *Quantiles of the Euler scheme for diffusion processes and financial applications*, Math. Finance **13** (2003), no. 1, 187–199.

203. Teman, E.: *Analysis of error with Malliavin Calculus: application to hedging*, Math. Finance **13** (2003), no. 1, 201–214.

204. Thalmaier, A., Wang, F.-Y.: *Gradient estimates for harmonic functions on regular domains in Riemannian manifolds*. J. Funct. Anal. **155** (1998) 109–124.

205. Üstünel, A.S.: *Une extension du calcul d'Itô via le calcul des variations stochastiques*, C. R. Acad. Sci. Paris Sér. I Math. **300** (1985), no. 9, 277–279.

206. _____: *La formule de changement de variables pour l'intégrale anticipante de Skorohod*, C. R. Acad. Sci. Paris Sér. I Math. **303** (1986), no. 7, 329–331.

207. _____: *An introduction to analysis on Wiener space*, Lecture Notes in Mathematics, vol. 1610, Springer-Verlag, Berlin, 1995.

208. Üstünel, A.S., Zakai, M.: *On the structure of independence on Wiener space*, J. Funct. Anal. **90** (1990), no. 1, 113–137.

209. _____: *Transformation of measure on Wiener space*, Springer Monographs in Mathematics, Springer-Verlag, Berlin, 2000.

210. Vargiolu, T.: *Invariant measures for the Musiela equation with deterministic diffusion term*, Finance Stoch. **3** (1999), no. 4, 483–492.

211. Villani, C.: *Topics in optimal transportation*, Graduate Studies in Mathematics, vol. 58, AMS, Providence, RI, 2003.

212. S. Watanabe, *Lectures on stochastic differential equations and Malliavin calculus*, Tata Institute of Fundamental Research Lectures on Mathematics and Physics, vol. 73, Published for the Tata Institute of Fundamental Research, Bombay, 1984.

213. _____: *Analysis of Wiener functionals (Malliavin calculus) and its applications to heat kernels*, Ann. Probab. **15** (1987), no. 1, 1–39.

214. Yoshida, N.: *Asymptotic expansion for statistics related to small diffusions*, J. Japan Statist. Soc. **22** (1992), no. 2, 139–159.

215. _____: *Asymptotic expansions of maximum likelihood estimators for small diffusions via the theory of Malliavin–Watanabe*, Probab. Theory Related Fields **92** (1992), no. 3, 275–311.

138   References

216. _____: *Malliavin calculus and asymptotic expansion for martingales*, Probab. Theory Related Fields **109** (1997), no. 3, 301–342.
217. _____: *Malliavin calculus and martingale expansion*, Bull. Sci. Math. **125** (2001), no. 6-7, 431–456.
218. Zakai, M.: *Malliavin derivatives and derivatives of functionals of the Wiener process with respect to a scale parameter*, Ann. Probab. **13** (1985), no. 2, 609–615.

Index